What Energy Sources Should Be Pursued?

Stuart A. Kallen, *Book Editor*

Bruce Glassman, *Vice President*
Bonnie Szumski, *Publisher*
Helen Cothran, *Managing Editor*

GREENHAVEN PRESS
An imprint of Thomson Gale, a part of The Thomson Corporation

D1498995

Detroit • New York • San Francisco • San Diego • New Haven, Conn.
Waterville, Maine • London • Munich

© 2005 Thomson Gale, a part of The Thomson Corporation.

Thomson and Star Logo are trademarks and Gale and Greenhaven Press are registered trademarks used herein under license.

For more information, contact
Greenhaven Press
27500 Drake Rd.
Farmington Hills, MI 48331-3535
Or you can visit our Internet site at http://www.gale.com

LIBRARY OF CONGRESS CATALOGING-IN-PUBLICATION DATA
What energy sources should be pursued? / Stuart A. Kallen, book editor.
p. cm. — (At issue)
Includes bibliographical references and index.
ISBN 0-7377-2757-8 (lib. : alk. paper) — ISBN 0-7377-2758-6 (pbk. : alk. paper)
1. Power resources. I. Kallen, Stuart A., 1955– . II. At issue (San Diego, Calif.)
TJ163.2.W45 2005
333.79—dc22   2004047479

Printed in the United States of America

Contents

Introduction

It would be difficult to imagine the modern world without cars, electric appliances, hot showers, air-conditioned buildings, and other modern amenities. Yet manufacturing and running all the machines that make life so easy and comfortable requires vast amounts of limited resources.

At the present time 89 percent of all the energy consumed in the United States is derived from fossil fuels—oil, coal, and natural gas. About 10 percent is derived from nuclear power and 1 percent from renewable sources such as the sun, wind, and water.

Although they are finite resources, fossil fuels have provided many obvious benefits to modern society. A small amount of gasoline, for example, provides a tremendous amount of energy for an automobile to travel hundreds of miles at high speeds. This gas is easy to transport with pipelines, ships, and trucks and is convenient and simple for consumers to use.

Reliance on oil comes with a price, however. While the United States uses fully 25 percent of the world's petroleum, it only has 5 percent of global petroleum reserves. Most of America's oil must be imported from foreign countries such as Saudi Arabia, Kuwait, Nigeria, and Venezuela at a cost to consumers of about $200,000 per minute. Besides creating a huge trade deficit, this situation requires Americans to rely on countries where brutal dictatorships, political instability, and outright hostility toward the United States can create serious oil shortages. As former CIA director James Woolsey notes, "Two-thirds of the world's oil reserves are in the Persian Gulf and as time goes on, the world is going to rely more on the Gulf, rather than less. . . . [You] can't bet on this part of the world making a smooth, easy path that will allow all of us to happily continue to drive our SUVs and use that part of the world as our filling station." Although sales of SUVs continue to climb, most Americans seem to understand Woolsey's point. A March 2003 Gallup poll reported that 65 percent of Americans believe the United States is likely to face a critical energy shortage during the next five years, up from 45 percent who expressed a similar fear in 1980.

Even if the gulf states remain content to keep America's wheels turning, the world currently consumes 29 billion barrels of oil every year. In the next fifteen years, that number is expected to nearly double even as production remains about the same. While the world will not run out of oil, as the years pass people will demand much more petroleum than the available supplies. This excessive demand will cause the price of a barrel of oil to rise dramatically, with potentially disastrous consequences for the world's economy. Major food shortages and war could result. According to energy industry expert Paul Roberts, writing in the *Los Angeles Times:*

> As production falls off . . . prices won't simply increase; they will fly. . . . [The] global economy is likely to slip into a recession so severe that the Great Depression [of the 1930s] will look like a dress rehearsal. Oil will cease to be a viable alternative—hardly an encouraging scenario in a world where oil currently provides 40% of all energy and nearly 90% of all transportation fuel. Political reaction would be desperate. . . . Worse, competition for remaining oil supplies would intensify, potentially leading to a new kind of political conflict; the energy war.

Even if supplies were to remain plentiful, oil poses other serious problems. Burning oil releases tons of pollution into the atmosphere every day. One of those pollutants, carbon dioxide (CO_2), is a gas that contributes to global warming. Other dangers to the environment, such as oil spills, wastewater contamination, and air pollution from refineries, also accompany petroleum use and production.

Coal is another fossil fuel that has great benefits and many drawbacks. While the fuel generates about 70 percent of the electricity in the United States, coal is a significant contributor to air and water pollution, acid rain, and global warming. Despite this situation, the nation's appetite for coal-powered electricity is expected to grow by 32 percent over the next twenty years. During this period, 27 percent of the aged, polluting coal-fired power plants now in operation will be worn out and scheduled for retirement.

Nuclear power plants do not contribute to air pollution, but they have other drawbacks. To produce energy, these power plants use uranium, a radioactive substance that is extremely poisonous. By storing this material, nuclear power plants have been shown to be vulnerable to terrorists. A small airplane flown into a building housing a nuclear reactor could contaminate an area the size of a large city for a period of thousands of years. In addition, the waste created by these reactors remains poisonous for tens of thousands of years. Scientists have yet to create a safe method for indefinite storage of spent nuclear fuel.

Despite the problems of oil, coal, and nuclear power, modern societies continue to benefit from the energy provided by these traditional sources. Looking to the future, however, many have called for a switch to renewable energy sources. As Glenn Hamer writes in *Power Engineering*, public support for these sources is growing for various reasons:

> People like to breathe clean air, and they care about the condition of the planet. Asthma rates are soaring, and when consumers who visit supermarkets learn that their fish is now dangerous to eat as a result of the mercury produced from coal generation, they naturally are going to want alternatives. This is not to mention the torrent of stories on global climate change, a condition that even the Bush Administration's EPA has acknowledged is at least partially caused by man.

Hamer is backed by a Gallup poll that shows 49 percent of Americans are concerned with protecting the environment. With that in mind, one of the major selling points for wind, solar, and hydrogen fuel cells are the environmental benefits of these new "green" energy sources. As former California Wind Energy coordinator Ty Cashman writes in *Earth Island Journal:*

> The wind blows in every country. The sun shines everywhere. . . . Local communities can have their own energy resources as well as their own local farms for food production. They will be able to produce their own electricity from wind and solar panels, and their own hydrogen fuel for homes, workshops and vehicles from their own water and solar energy.

One of the most promising forms of renewable energy is solar power. Solar power utilizes the enormous amount of energy produced by the sun every day. Solar collectors work in a variety of ways to turn this energy into heat and electricity. According to some studies, enough electrical power could be generated in a one-hundred-square-mile area of the desert southwest to supply the entire country with electricity. Demand for solar power energy is growing. The U.S. market for solar power increased approximately 60 percent in 2003, and global production has grown sixfold since 1997. More than a million Americans use solar water heaters, and more than two hundred thousand homes use photovoltaic systems.

Like all energy sources, however, there are drawbacks to solar power. For example, at current rates the investment needed to produce solar collectors makes the power more expensive than electricity generated by fossil fuels. In addition, there is no way to efficiently store large amounts of solar power for use at night and for prolonged cloudy periods.

Another source of renewable energy—the hydrogen fuel cell—is being developed to provide power for automobiles. These cells separate hydrogen molecules from oxygen in water in order to produce electricity. Backed by grants from the federal government, major auto companies such as DaimlerChrysler, Ford, and General Motors are spending billions of dollars to develop practical applications for fuel cells within the next decade. According to hydrogen researcher and author Seth Dunn, "The critical question is no longer whether we are headed toward hydrogen, but how we should get there, and how long it will take."

Like other power sources, however, hydrogen has its critics. Although there are substantial commercial, political, and environmental benefits to developing fuel cells, the government is planning on using large amounts of a fossil fuel—natural gas—to create hydrogen. While hydrogen can be produced by solar power, at the present time there is no infrastructure to provide the vast amounts of hydrogen that would be necessary to keep America's 90 million cars and trucks on the road.

The advantages and disadvantages of solar energy and hydrogen fuel cells are among the issues debated in *At Issue: What Energy Sources Should Be Pursued?* Throughout this anthology, energy industry experts, environmentalists, and others disagree as to the best path to follow, but most are convinced that the current energy system is not going to function in the next century the way it has in the past. All agree that renewable energy, more efficient oil production techniques, or technologies yet to be discovered must be developed in order to meet the energy needs of the future.

1

The World's Oil Supply Is Plentiful

David Deming

David Deming is an adjunct scholar with the National Center for Policy Analysis and associate professor of geology and geophysics at the University of Oklahoma.

While some proclaim that the world's oil supply is dwindling, new technology is allowing companies, for the first time, to recover oil hidden deep in the earth. In addition, new methods for extracting petroleum from tar and shale ensure that the United States will never run out of oil. In fact, a new modern age fueled by petroleum is just beginning.

It is difficult to imagine how our grandparents and great-grandparents lived at the end of the nineteenth century. The United States was still largely a rural society, and the amenities we take for granted today were unknown then.

Most people lived on farms. Few Americans had running water, hot water, bath tubs, or flush toilets. Central heating, electricity, and telephones were rare. There were no antibiotics. Infant mortality was high, and life expectancy was 30 years lower than it is today. For most people, educational opportunities were very limited. In 1890, only 5 percent of the eligible population attended high school.

In the year 1900, there were only about 8,000 automobiles in the entire country. Horseless carriages, like yachts, were a toy for the rich to enjoy. People knew there would never be enough gasoline to power a nation of automobiles because the output

David Deming, "Abundant Reserves Show Petroleum Age Is Just Beginning," Heartland Institute, www.heartland.org, October 1, 2003. Copyright © 2003 by David Deming. Reproduced by permission.

of the Pennsylvanian oil fields had been declining for years.

The seminal event that transformed the United States into an industrial and technological powerhouse occurred on the morning of January 10, 1901, near Beaumont, Texas. A wildcat oil well on a location named Spindletop erupted into a geyser 100 feet high. It was the greatest oil well ever seen in the United States.

> *Every year, technological advances make it possible to draw upon petroleum resources whose extraction was once unthinkable.*

Over the next year, production from the Spindletop well equaled the production of 37,000 typical oil wells in the eastern U.S. Overnight, the price of oil dropped to 3 cents a barrel, recovering to 83 cents a barrel two-and-a-half years later. The cheap energy provided by abundant oil allowed the U.S. to transform itself from a rural, agrarian country into an urban, industrialized nation. Along the way, the prosperity of our society increased manyfold.

Petroleum continues to be the lifeblood of our civilization. In the U.S., our entire way of life depends on the energy provided by the oil industry. Oil and natural gas are by far the most important energy sources for the world.

Technology prevails

And the age of petroleum has only just begun. For more than 80 years, geologists' estimates of the world's endowment of oil have risen faster than developers can pump it out of the ground. In 1920, the U.S. Geological Survey estimated just 20 billion barrels of oil remained in the world. By the year 2000, the estimate had grown to 3,000 billion barrels.

Every year, technological advances make it possible to draw upon petroleum resources whose extraction was once unthinkable. We can now drill wells up to 30,000 feet deep. The amount of oil that can be recovered from a single well has been enhanced by a technology that allows multiple horizontal shafts to be branched off from one vertical borehole. The ability to drill offshore in water depths of up to 9,000 feet has opened up the vast petroleum resources of the world's sub-merged continental margins.

The world also contains immense amounts of unconventional oil resources that we have not yet begun to tap. Tar sands found in Canada and South America contain 600 billion barrels of oil, enough to supply the U.S. with 84 years of oil at the current consumption rate. Worldwide, the amount of oil that can be extracted from oil shales could be as much as 14,000 billion barrels—enough to supply the world for 500 years.

Oil is cheap, abundant, and clean. Dramatic improvements in our quality of life over the past two centuries can be traced back to the energy provided by the petroleum industry. Without that energy, and the technological advances that make its recovery possible, life in the U.S. would be short, dark, and impoverished.

2

The World Is Running Out of Oil

Stuart H. Rodman

Stuart H. Rodman is the author of The Last Days of Power?
*and an alternative energy expert whose reports have been
heard on the nationally syndicated radio show* Coast to
Coast AM *with Art Bell, and on the Internet show* Sightings
on the Radio with Jeff Rense.

While there will always be crude oil in the ground, the
energy expended drilling and transporting it will soon
nearly equal the energy provided by the oil itself. When
this happens, prices will skyrocket and the petroleum
age will come to an end forever. Despite optimistic government reports and improved drilling technology, this
day of reckoning will be within the lifetimes of most
people alive today. Dwindling oil supplies will also
make it difficult and expensive to pursue alternative energy sources.

You're probably thinking . . . that there is really plenty of
cheap gas to go around. Why worry? But the mother of all
oil shocks could be just 5 or 6 years away! . . .

Let's hope not, because our current lifestyles are dependent
on oil for everything from manufacturing, to transportation, to
agriculture. Despite this and even in the face of the recurrent
oil shocks of the last decades, very little has been done worldwide to lessen our addiction to the "black gold" from within.

And consider this. Regardless of how much petroleum re-
sides in the bowels of the earth, when the production of a
given amount of fuel requires the industry to first consume the

equivalent amount to discover, extract, refine, and deliver it, it's all over. You might as well be out of gas. But guess what? Despite advances in technology, that day may be much closer than you think. What's more, when the amount of energy produced per barrel of oil is just equal to the amount used to obtain it, nothing will be left to run the engines of commerce or agriculture either.

How could that be? Suppose you wanted to drive to the nearby filling station to buy some gas and you only had a gallon in your tank. What if you suddenly remembered that you would have to use all your gas to get there though and that you had been told that the station would only allow you to buy just one gallon. You would have enough gas then to get from the station back to where you started from but no more. If you have to use all your fuel just to get an equivalent amount there would be no point leaving home in the first place.

The issue is known as "net energy" and it may be more important to understanding our future than worrying about . . . all the oil available for future extraction from our planet. Here's why.

The end of fossil energy

Speaking of net energy, oil industry watcher Jay Hanson states, citing the laws of thermodynamics, "By definition, energy 'sources' must generate more energy than they consume; otherwise, they are 'sinks'."

But net-energy analysis first reached public attention in 1974. At that time, *Business Week* reported that oil scientist Howard Odum had developed a "New Math for Figuring Energy Costs." To the surprise of many, Odum's new math indicated that [some] oil well operations were energy sinks and not energy sources.

According to this analysis, these operations could be profitable only when "subsidized" by cheap, regulated oil, which was used to produce deregulated oil. Because the industry is subsidized in this way in terms of net energy and also from direct taxpayer "allowances", the industry can continue to produce oil at a monetary profit, at least for a while. Hanson observes, "Even without direct and indirect subsidies of $650 billion a year it's conceivable that energy companies could make money—but lose energy—by burning one $10-barrel of oil today in order to pump one-half of a $50-barrel tomorrow."

But how much longer can they keep this up? Hanson says, "Based on the best information we have at hand today, sometime during . . . [the 21st century] the global economy will 'run out of gas', as fossil energy sources become sinks. One can argue about the exact date this will occur, but the end of fossil energy—and the dependent global economy—is inevitable."

> **When oil production peaks worldwide, most experts agree it will be a whole new ballgame. That day of reckoning is inevitable.**

Of course major oil companies may have private reserves of fuel which can be used to underwrite the energy costs of future production. However, if the energy produced for distribution to society is not sufficient to also pay back the overhead, the reserves themselves will eventually evaporate.

And those in denial might think that oil supplies will last forever. In reality though of course, the oil supply is finite. Jim Bell, author of *Achieving Economic Survival on Spaceship Earth*, compares oil exploration with picking apples from a tree, "We tend to pick the ones that have fallen from the tree and those that are closest to the ground first. Later though we may need a ladder as we expend more effort to find the ones that are harder to reach. The oil companies have to try harder and harder each year to find extractable oil. They have already harvested the easiest pickings."

Jim Bell points out, "We are always told that there is plenty of ultimately recoverable oil left in the ground and so we naturally assume the supply will last forever, certainly through our lifetimes. What really matters though is not the amount of petroleum that lies within our planet, but the price we pay for it, in terms of our wallets and the consequences for our planet. And higher fuel prices will raise the price of everything else. As the production of petroleum begins to decline, market forces will push the price of hydrocarbon based products, if you can find them at all, higher and higher."

So then in the pursuit of profit, as the oil companies expend more and more energy to extract and refine petroleum, they eventually reach the point of diminishing returns, as proven reserves are depleted. Although there may be more ultimately recoverable crude in the ground, the new sources tend most often to be smaller, require more energy expenditure, or are technologically unexploitable. At that point, production "peaks", then declines rapidly.

Cheap oil gone forever

When oil production peaks worldwide, most experts agree it will be a whole new ballgame. That day of reckoning is inevitable. But when? It is a known fact that oil production peaked in this country in the 1970s and impressive evidence suggests that oil production worldwide will peak during the next twenty five years or sooner, some say as early as 2010. Hanson states that "the petroleum industry itself has announced that global oil production will 'peak' in less than ten years!"

You wouldn't know that from the official reports however. According to the petroleum industry's own spokespersons, there is at least another 93 years of known petroleum reserves worldwide to keep us in gas, at the current rate of consumption. The U.S. government has been even more optimistic. Past government studies cite advances in technology and the promise of synthetic fuels and methodologies as being cause to expect the continued availability of petroleum based fuels for generations to come. More recent reports though by the United States Geological Survey offer cautions about the coming end of the "buyer's market" for petroleum in a world of growing demand.

> **❝** *'Energy resources that consume more energy than they produce are worthless as sources of energy.'* **❞**

However, when the peak comes, whenever it does, all bets are off. And, it could be preceded by serious production shortages, which could occur even sooner. But Hanson warns the peak will come sooner and not later. When it does, Hanson states, "The price of oil is expected to rise sharply—and permanently."

And he has good reason to say so. In another paper, "The Best Kept Secret in Washington", Hanson discusses a private study that was conducted by worldwide industry expert, "Petroconsultants" (now known as IHS Group). The study suggests that when the peak comes, the markets will treat petroleum as a scarcity and that the days of "cheap" petroleum will then be gone forever. . . .

And they are not the only experts sounding the alarm. In a recently published report from *The New Republic*, Gregg Easterbrook reports that highly respected industry analyst Colin Campbell holds similar views: "Campbell bases his thinking on something called the Hubbert Curve, perfected by M. King Hubbert, patron saint of petroleum geologists. Hubbert found that production tends to peak almost exactly when a petroleum reservoir hits its halfway point—meaning that once a well's output begins to decline, the amount left in the ground is roughly equal to what has been pumped out. In 1956, when oil optimism was universal, Hubbert used his curve to forecast that U.S. petroleum production would peak in 1969. The actual peak came in 1970; this dead-on prediction has given Hubbert legendary status."

> *'Obviously the death sentence for billions of people has already been issued.'*

Easterbrook goes on to say, "The evidence is legion. In the United States, which contains 75 percent of the world's oil wells, petroleum production has been in decline since the 1970 peak. [Alaska's] Prudhoe Bay, the last 'elephant' oil find in the United States, peaked in 1988. Production in the former Soviet states also peaked that year."

Some experts say in fact that there are nearly 500,000 wells sites in the U.S. that produce less than a single barrel of oil per day. An added exclamation comes from Campbell's own work with [French petroleum expert Jean] Luherre: "By 2002 or so the world will rely on Middle East nations, particularly five near the Persian Gulf (Iran, Iraq, Kuwait, Saudi Arabia and the United Arab Emirates), to fill in the gap between dwindling supply and growing demand. But once approximately 900 gbo (900 thousand billion barrels of oil) have been consumed, production must soon begin to fall. Barring a global recession, it seems most likely that world production of conventional oil will peak during the first decade of the 21st century."

And after the peak? Campbell says, "From an economic

perspective, when the world runs completely out of oil is thus not directly relevant: what matters is when production begins to taper off. Beyond that point, prices will rise unless demand declines commensurately."

Sinking ship

A nonrenewable resource, the end of cheap oil is inevitable, although some may choose to argue the timeline. Debate aside, the next great oil shock, however, may have nothing to do with money or supply and everything to do with that other problem, the one no one wants to talk about, net energy. Hanson notes, "The key to understanding energy issues is to look at the 'energy price' of energy. Energy resources that consume more energy than they produce are worthless as sources of energy. This thermodynamic law applies no matter how high the 'money price' of energy goes. For example, if it takes more energy to search for and mine a barrel of oil than the energy recovered, then it makes no energy sense to look for that barrel—no matter how high the money price of oil goes."

Consider this illustration from University of Wisconsin at Stevens Point professor Thomas Detwyler: "The useful energy to be obtained from nonrenewable resources, such as fossil fuels (mainly crude oil, coal and natural gas) and uranium, is subject to diminishing returns through time. It takes energy to get energy. And because we exploit the easiest-to-get energy resources first, each subsequent unit of gross energy (e.g., oil in the ground) requires greater energy subsidy to obtain than did the previous unit, thus leaving less net energy."

No doubt. Energy costs in the oil industry are on the rise and are reflected in the increasing depth of wells: 300 feet in 1870, 1,000 feet in 1900, 3,000 feet in the 1920s and more than 6,000 feet by 1980. Campbell notes, "The cost of drilling oil and gas wells (which is largely a function of energy subsidy) rises exponentially with increasing depth. By the mid-1970s, about half the petroleum produced in Texas was also consumed there as production-related subsidies, so that at best net energy was only half of gross energy."

It is for this reason perhaps that net energy returns have been falling consistently despite improved technology. Campbell adds, "The dynamic of shrinking net energy means that the usefulness of gross energy reserves may be vastly overrated. In fact, a large portion of any given gross reserve will be energetically unexploitable though perhaps technically extractable. Beyond the resource cutoff line, the system is an energy sink requiring more energy as subsidy than is returned as net energy."

Just how bad is it? Citing work by analyst John Gever from "Beyond Oil", Hanson states that in the '50s the industry could produce 50 barrels of energy for every barrel consumed producing finished products for the market. By the nineties, the ration had fallen to 5 barrels to 1. By the year 2005, the industry will just break even—it will be necessary to use as much energy to produce any given quantity.

Hanson adds, "Under that latter scenario, even if the price of oil reaches $500 a barrel, it wouldn't be logical to look for new oil in the US because it would consume more energy than it would recover."

Difficult to pursue alternatives

It takes energy to make energy. As supplies shrink and prices rise, market forces may drive us towards other fuel sources. But we will need to have an infrastructure in place capable of assuring an uninterruptible supply. Professor Robert Costanza of the University of Maryland cautioned though that there is an "embodied cost" of energy, whereby manufactured goods, like machinery, power cable, relays, switchboxes, or any finished goods, exist only after a given amount of energy was consumed by industry in their manufacture.

And Jim Bell explains, "When the amount of net energy available in society begins to shrink it is harder to harness the resources necessary to manufacture the solar panels, the wind mills, and the other equipment needed when we begin the inevitable task of creating a large scale alternative infrastructure."

In reality, because there is only a dwindling supply of energy that can be sucked from the well, absent an alternative, we will be living in an "energy limited economy". Hanson offers this definition: "An 'energy-limited economy' is one where more energy cannot be had at any price. The global economy will become 'energy-limited' once global oil production peaks in less than ten years (perhaps much less)."

That could mean more trouble than just lining up at the filling station. Consider the problem facing agriculture. Hanson points out, "Food grains produced with modern, high-yield methods (including packaging and delivery) now contain between four and ten calories of fossil fuel for every calorie of solar energy." Hanson adds, "It has been estimated that about four percent of the nation's energy budget is used to grow food, while about 10 to 13 percent is needed to put it on our plates. In other words, a staggering total of 17 percent of America's energy budget is consumed by agriculture!" Again citing other sources, Hanson states, "By 2040, we would need to triple the global food supply in order to meet the basic food needs of the eleven billion people who are expected to be alive. *But* doing so would require a 1,000 percent increase in the total energy expended in food production."

Following the peak of oil production, absent an alternative, Hanson notes, "It will be physically impossible—*thus economically impossible*— to provide enough net energy to agriculture."

Hanson adds grimly, "Obviously the death sentence for billions of people has already been issued."

So what can be done? Conservation, building a better light bulb or even just a more fuel efficient SUV, though a good idea, won't be enough. Observers like Bell and others suggest that unless industry recognizes the need to shift now from a mind set that views energy resources in terms only of dollars instead of in terms of diminishing returns in real energy units, they will be planting the seeds of their own destruction, simply throwing good money after bad. Hanson notes, "To have more energy in the future means that energy must be diverted now from non-energy sectors of the economy into future energy generation."

But what about the so called "unconventional oil"—the oil found in shale deposits and sand tars? Sinks all. In just a few short years then, the net energy value of oil could be zero and its fate as the world's dominant energy source will be sealed. Light's out! When civilization can only produce the amount of energy needed to cover the energy expended to produce itself, nothing is left to power your car, run your business, or even grow you food. At that point, the size of the world's crude reserves would be irrelevant. Oil would then become of greater interest to historians than to consumers as the Age of Petroleum joins the Bronze Age and the Stone Age as footnotes in the chronicling of civilizations.

3

Alternatives to Oil Must Be Developed

Walter Youngquist

Walter Youngquist is a retired professor of geology at the University of Oregon who develops scenarios of future oil production. He is the author of GeoDestinies: The Inevitable Control of Earth Resources over Nations and Individuals.

Petroleum fuels the modern world. While many alternative energy sources exist, few can compare with oil's availability, convenience, and versatility. Unfortunately, petroleum is a finite resource; in the near future the world will run out of cheap, easily recoverable fuel. While there are fifteen major alternatives to oil, all have disadvantages when it comes to running cars, trucks, and electrical power plants. Despite these shortcomings, scientists, researchers, and world leaders must find a way to harness alternative energy while slowing world population growth. Otherwise, future economic and humanitarian disasters are inevitable.

O il fuels the modern world. No other substance can equal the enormous impact which the use of oil has had on so many people, so rapidly, in so many ways, and in so many places around the world.

Oil in its various refined derivative forms, such as gasoline, kerosene, and diesel fuel, has a unique combination of many desirable and useful characteristics. These include a current availability in abundance, a currently high net energy recovery, a high energy density, ease of transportation and storage, relative safety, and great versatility in end use. Oil is also useful as more than an energy source. It is the basis for the manufacture of petrochemical products including plastics, medicines, paints, and myriad other useful materials. Finally, the asphalt "bottoms" from refineries have [been used as paving materials to convert] millions of miles of muddy trails around the world into paved highways on which transport vehicles fueled by oil run.

Alternative energy sources must be compared with oil in all these various attributes when their substitution for oil is considered. None appears to completely equal oil.

But oil, like other fossil fuels, is a finite resource. True, there will always be oil in the Earth, but eventually the cost to recover what remains will be beyond the value of the oil. Also, a time will be reached when the amount of energy needed to recover the oil equals or exceeds the energy in the recovered oil, at which point oil production becomes no more than a break-even, or a net energy loss situation.

Oil being the most important of our fuels today, the term "alternative energy" is commonly taken to mean all other energy sources and is used here in that context. Realizing that oil is finite in practical terms, there is increasing attention given to what alternative energy sources are available to replace oil. The imperative to pursue alternative energy sources is clearly established by two simple facts. The world now uses more than 26 billion barrels of oil a year, but new discoveries (not existing field additions) in recent years have been averaging less than seven billion barrels yearly. The peak of world oil discoveries was in the mid-1960's. Inevitably, the time of the peak of world oil production must follow, with most current estimates ranging from the year 2003 to 2020. Significantly, all estimates of production peak dates are within the lifetimes of most people living today.

The amount of energy an individual can directly or indirectly command largely determines that individual's material standard of living. This, of course, also applies to nations as a whole. To provide adequate energy for future generations introduces the concept of sustainability. What significant energy sources can be drawn on indefinitely?

Problems of growth and energy

"Sustainable" is a popular and pleasant word, but when it is used it needs to be clearly defined and placed within certain parameters. The term "sustainable growth" is popular with Chambers of Commerce as well as with corporations, but if this means increase in use of any resource, including land for more people, more water for more people, and more and more food, or more "things", then the term "sustainable growth" is an oxymoron. Growth in terms of numbers of anything cannot be sustained indefinitely. Sustainable growth in terms of better medical care, improved sanitation, and other related qualities of life, and of intellectual endeavors, among other things, is possible, and should be a continual goal.

> *The transition to alternative fuels will not be simple nor as convenient as is the use of oil today, and it will involve much time and financial investment.*

Any consideration of "sustainable" must also be framed in the concept of a fixed size of population. People use resources. And all energy resources, even solar energy, are limited. The problem of population size is politically sensitive and therefore largely avoided in discussions. But the energy problem cannot be sustainably solved if the demand target is a continually growing population. It is important to keep this overriding fact in mind. Eventually it will have to be faced. In defining a sustainable society, it is also necessary to determine what a reasonable standard of living is to be achieved. This does not lend itself to an easy definition as various cultures have differing views.

In considering what significant (in terms of quantity and quality) sustainable alternative energy sources may exist, the factors of population and living standards must be addressed. These matters are beyond this discussion, which simply presents the basic facts of alternative energy sources. How these sources, with their advantages and limitations may be applied to society at large is here left for economists, sociologists, and politicians. . . .

There is much casual popular thought that energy sources are easily interchangeable. "When we run out of oil we will go to alternative fuels." "We can run our cars on solar energy." Such statements are legion. But the transition to alternative fuels will not be simple nor as convenient as is the use of oil today, and it will involve much time and financial investment. Energy carriers, in terms of varied end uses and ease of handling and storage, are not easily interchangeable.

We here briefly examine alternative energy sources as to their advantages, limitations, and their prospects for replacing oil in the ways and great volumes in which we use oil today. Alternative energies closest to conventional oil (from wells) are first considered, and then our energy horizons are expanded.

Energy sources can be divided into renewable and nonrenewable.

Nonrenewable energy sources

Oil sands/heavy oil. This oil exists in huge quantities (trillions of barrels) particularly in Alberta, Canada and Venezuela. It is true oil but in deposits which take special methods to recover the oil. Oil sands must either be mined, or recovered by the SAGD (SAG-D) process (steam assisted gravity drainage) in which steam is injected in the upper of two parallel pipes and the oil is collected in the lower pipe. The oil must have lighter hydrocarbons added to it to allow it to flow and be processed into conventional petroleum products. Heavy oil deposits can be injected with hot water or steam. Because of the energy expended in these processes, the net energy recovery is considerably less than oil from conventional drilled wells.

> **//** *There is no battery pack which can effectively move heavy farm machinery over miles of farm fields, and no electric battery system seems even remotely able to propel a Boeing 747.* **//**

At present about 500,000 barrels a day are recovered from the Athabasca oil sands of Alberta. To increase this 10-fold to 5 million barrels a day would be a very large task, with severe environmental limitations. This must be put in the perspective of the 76 million barrels of oil the world now consumes daily. Other similar oil deposits have the same problems of scale and net energy recovery. In total, oil sands and heavy oil can replace conventional oil only to a small degree. Canada's domestic needs for oil, with its growing population and increasing industrialization, will likely soon absorb all the additional oil which can be produced from oil sands and heavy oil with no surplus to export.

Natural gas. Natural gas is methane (CH4), which commonly has minor quantities of noncombustible gases such as carbon dioxide and nitrogen associated with it. Natural gas is termed "associated gas" when it occurs with oil, or "nonassociated gas" when it is not found with oil. Natural gas is derived from organic material and can be formed at essentially normal atmospheric temperature (such is the origin of "swamp gas" and the gas associated with garbage dumps, now in places used for fuel to generate electric power). . . .

Natural gas is the cleanest burning of the fossil fuels, and for that reason is the fuel of choice over coal for electricity production as boiler fuel and in gas turbines. Natural gas can be used as a substitute for gasoline or diesel fuel in internal combustion engines, and is so used in a few places.

Natural gas is commonly moved by pipeline. It can be shipped in cryogenic tankers but this is expensive and does not lend itself economically to large scale transport, whereas oil is shipped economically worldwide. Natural gas can be converted to a liquid (GTL—gas to liquid), and such conversion plants are being built in areas not served by pipelines (e.g., the North Dome Field of Qatar). The end product is a high grade substitute for gasoline. However, the volumes of GTL which can be produced are modest and somewhat more expensive than gasoline.

Natural gas is more widely distributed than oil. But estimates are that in total its energy in reserves is equal to or slightly less than that in world oil reserves. Natural gas (and in GTL) is an alternative energy to petroleum, but natural gas is also a finite fossil fuel.

Mining and processing

Coal. Coal is a very large energy source, but it must be mined, it is not nearly so easy to handle and transport as is oil, and it has much less energy density. For use in producing electricity in power plants (burned under boilers), coal can replace oil. But converting it to a liquid fuel which might be used in motor vehicles is expensive, and doing this on a scale which could significantly replace oil in vehicle use would require impossibly large mining projects. Coal can replace oil in some uses. Although considerable progress has been made, coal production and burning still have environmental problems which are of major concern. Adding to the greenhouse effect is one. The energy in coal reserves worldwide is greater than oil, but it, too, is a finite fossil fuel.

Shale oil. Production of oil from oil shale has been attempted at various times for nearly 100 years. So far, no venture has proved successful on a significantly large scale. One problem is that there is no oil in oil shale. It is a material called kerogen. The shale has to be mined, transported, heated to about 4500C (8500F), and have hydrogen added to the product to make it flow. The shale pops like popcorn when heated so the resulting volume of shale after the kerogen is taken out is larger than when it was first mined. The waste disposal problem is large. Net energy recovery is low at best. It also takes several barrels of water to produce one barrel of oil. The largest shale oil deposits in the world are in the Colorado Plateau, a markedly water poor region. So far shale oil is, as the saying goes: "The fuel of the future and always will be." B. Fleay states: "Shale oil is like a mirage that retreats as it is approached." Shale oil will not replace oil. . . .

Nuclear power

Although uranium . . . is a finite resource, converting uranium-238 to plutonium-239 (a process called "breeding") could possibly extend our use of uranium for [nuclear] power by perhaps 100 times. However, plutonium is an exceedingly toxic substance, and also the basis for a deadly bomb. Because of this there is much opposition to the breeder reactor, and to uranium for power in general due to safety and environmental considerations. However, coal and uranium are the only two alternative sources of energy which can be developed in large amounts, and provide a dependable base load in the reasonably near future. Nuclear power development has been stopped in the United States. Elsewhere, some countries are abandoning nuclear power (e.g., Sweden, Germany), whereas others are pursuing it (e.g., Japan, Russia). Ultimately, however, nuclear power in any form is nonrenewable because uranium reserves are limited.

The end product of nuclear fission is electricity. How to use electricity to efficiently replace oil (gasoline, diesel, kerosene) in the more than 700 million vehicles worldwide has not yet been satisfactorily solved. There are severe limitations of the storage batteries involved. For example, a gallon of gasoline weighing about 8 pounds has the same energy as one ton of conventional lead-acid storage batteries. Fifteen gallons of gasoline in a car's tank are the energy equal of 15 tons of storage batteries. Even if much improved storage batteries were devised, they cannot compete with gasoline or diesel fuel in energy density. Also, storage batteries become almost useless in very cold weather, storage capacity is limited, and batteries need to be replaced after a few years use at large cost. There is no battery pack which can effectively move heavy farm machinery over miles of farm fields, and no electric battery system seems even remotely able to propel a Boeing 747 14 hours nonstop at 600 miles an hour from New York to Cape Town (now the longest scheduled plane flight). Also, the considerable additional weight to any vehicle using batteries is a severe handicap in itself. In transport machines, electricity is not a good replacement for oil. This is a limitation in the use of alternative sources where electricity is the end product.

> *Direct conversion of sunlight to electricity by solar cells is a promising technology . . . , but the amount of electricity which can be generated by that method is not great compared with demand.*

Where oil is used for electric power production, nuclear fission can replace oil as a fuel. However, in the U.S. now only about 2 percent of electric power is generated from oil. Elsewhere, such as island economies, oil is now the chief source for electric power generation and nuclear fission has the prospect of significantly replacing that oil.

Geothermal energy

This is heat from the Earth. In a few places in the world there is steam or very hot water close enough to the surface so that the resource can be reached economically with a drill. The steam, or hot water flashed to steam, can turn a turbine, turning a generator producing electricity. At best, because of the scarcity of such sites, geothermal energy can be only a minor contributor to world energy supplies, and the product is electricity, which is subject to limited end uses. It should be noted that all electric power geothermal generating site reservoirs are now declining, because the geothermal requirements to produce electric power draw down the reservoirs faster than their recharge ability. Some projects are now reinjecting water from the condensed steam back into the reservoir to see if this problem can be mitigated, but results so far are inconclusive. However, when lower temperature reservoirs are used for space heating, with a more modest demand on the reservoir using down-well heat exchangers or ground to air heat pumps using the natural heat flow of the Earth, geothermal energy appears to be a renewable energy source.

Renewable energy sources

Wood and other biomass. Wood has long been used as a fuel, now to the extent that large areas worldwide are being deforested resulting in massive erosion in such places as the foothills of the Himalayas, and the mountains of Haiti. Wood can be converted to a liquid fuel but the net energy recovery is low, and there is not enough wood available to be able to convert it to a liquid fuel in any significant quantities.

Other biomass fuel sources have been tried. Crops such as corn are converted to alcohol. In the case of corn to ethanol, it is an energy negative. It takes more energy to produce ethanol than is obtained from it. Also, using grain such as corn for fuel, precludes it from being used as food for humans or livestock. It is also hard on the land. In U.S. corn production, soil erodes some 20-times faster than soil is formed. Ethanol has less energy per volume than does gasoline, so when used as a 10 percent mix with gasoline (called gasohol), more gasohol has to be purchased to make up the difference. Also, ethanol is not so environmentally friendly as advocates would like to believe. [Professor of environmental studies David] Pimentel states:

> Ethanol produces less carbon monoxide than gasoline, but it produces just as much nitrous oxides as gasoline. In addition, ethanol adds aldehydes and alcohol to the atmosphere, all of which are carcinogenic. When all air pollutants associated with the entire ethanol system are measured, ethanol production is found to contribute to major air pollution problems.

With a lower energy density than gasoline, and adding the energy cost of the fertilizer (made chiefly from natural gas), and the energy costs (gasoline and/or diesel) to plow, plant, cultivate, and transport the corn for ethanol production, ethanol in total does not save fossil fuel energy nor does its use reduce atmospheric pollution.

> *Fuel cells are not a source of energy in themselves, but are a possible ultimate substitute for the internal combustion engine.*

A comprehensive study of converting biomass to liquid fuels by [author Mario] Giampietro and others conclude:

> Large scale biofuel production is not an alternative to the current use of oil, and is not even an advisable option to cover a significant fraction of it.

Hydro-electric power. Originally thought of as a clean, non-polluting, environmentally friendly source of energy, experience is proving otherwise. Valuable lowlands, which are usually the best farmland, are flooded. Wildlife is displaced. . . . The effect on fish has been disastrous. Only to a small extent is hydro-electric power truly renewable. . . . If reservoirs are involved, in order to provide a dependable base load as is the case of most hydro-electric facilities, hydro-electric power in the longer term is not a truly renewable energy source. All reservoirs eventually fill with sediment. Some reservoirs have already filled, and many others are filling faster than expected. A dam site can be used only once.

We are enjoying the best part of the life of huge dams. In a few hundred years Glen Canyon Dam and Hoover Dam will be concrete waterfalls. And, again, the end product is electricity, not a replacement for the important use of oil derivatives (gasoline, etc.) in transportation equipment.

Sun, wind, and waves

Solar energy. This is a favorite possible source of future energy for many people, comforted by the thought that it is unlimited. But, quite the contrary is true. The Sun will exist for a long time, but at any given place on the Earth's surface the amount of sunlight received is limited—only so much is received. And at night, or with overcast skies, or in high latitudes where winter days are short and for months there may be no daylight at all, or available in small and low intensity quantities. Direct conversion of sunlight to electricity by solar cells is a promising technology, and already locally useful, but the amount of electricity which can be generated by that method is not great compared with demand. Because it is a low grade energy, with a low conversion efficiency (about 15%) capturing solar energy in quantity requires huge installations—many square miles. About 8 percent of the cells must be replaced each year. But the big problem is how to store significant amounts of electricity when the Sun is not available to produce, for example, at night. The problem remains unsolved. Because of this, solar energy cannot be used as a dependable base load. And, the immediate end product is electricity, a very limited replacement for oil. Also, adding in all the energy costs of the production and maintenance of PV (photovoltaic) installations, the net energy recovery is low.

> *A sustainable world order must be based on acceptance of much lower per capita levels of energy use, much lower living standards and a zero growth economy.*

Wind energy. This energy source is similar to solar in that it is not dependable. It is noisy, and the visual effects are not usually regarded as pleasing. The best inland wind farm sites tend to be where air funnels through passes in the hills which are also commonly flyways for birds. The bird kills have caused the Audubon Society to file suit in some areas to prevent wind energy installations. Locally and even regionally via a grid (e.g., Denmark) wind can be a significant electric power source. But wind is likely to be only a modest help in the total world energy supply, and the end product is electricity, no significant replacement for oil. As with solar energy, the storage problem of large amounts of wind generated electricity is largely unsolved. Wind cannot provide a base load as winds are unreliable.

Wave energy. All sorts of installations have been tried to obtain energy from this source, but with very modest results. Piston arrangements moved up and down by waves which in turn move turbines connected to electric generators have been tried in The Netherlands, but the project was abandoned. Waves are not dependable, and the end product is electricity, and producing it in significant quantities from waves seems a remote prospect.

Tidal power. It takes a high tide and special configuration of the coastline, a narrow estuary which can be dammed, to be a tidal power site of value. Only about nine viable sites have been identified in the world. Two are now in use (Russia and France) and generate some electricity. Damming estuaries would have considerable environmental impact. The Bay of Fundy in eastern Canada has long been considered for a tidal power site, but developing it would have a negative effect on the fisheries and other sea-related economic enterprises. It would also disturb the habits of millions of birds which use the Bay of Fundy area as part of their migration routes. Tidal power is not a significant power source. The end product is electricity.

Fusion. Fusion involves the fusion of either of two hydrogen isotopes, deuterium or tritium. Deuterium exists in great quantities in ordinary water, and from that perspective fusion is theoretically an almost infinitely renewable energy resource. This is the holy grail of ultimate energy. Fusion is the energy which powers the Sun, and that is the problem. The temperature of the Sun ranges from about 100,000C on its surface to an estimated 15 to 18 million degrees in the interior where fusion takes place. Containing such a temperature on Earth in a sustainable way and harnessing the heat to somehow produce power has so far escaped the very best scientific talent. However, even if commercial fusion were accomplished, the end product again is electricity, not a direct convenient replacement for oil.

> *We have done remarkably little to reduce our dependence on a fuel [oil] which is a limited resource and for which there is no comprehensive substitute in prospect.*

Ocean thermal energy conversion (OTEC). Within about 25 degrees each side of the equator the surface of the ocean is warm, and the depths are cold to the extent that there is a modest temperature differential. This can be a source of energy, using a low boiling point fluid such as ammonia which at normal atmospheric temperature of 700F (210C) is a gas, colder water can be pumped from the deep ocean to condense the ammonia, and then let it warm up and expand to gas. The resulting gas pressure can power a turbine to turn a generator. But the plant would have to be huge and anchored in the deep open ocean or on a ship, all subject to storms and corrosion, and the amount of water which has to be moved is enormous as the efficiency is very low. How to store and transport the resulting electricity would also be a large problem. OTEC does not appear to have much potential as a significant energy source, and the end product is electricity.

Hydrogen and fuel cells

References are sometimes made as to using these for energy sources. Neither is a primary energy source. Hydrogen must be obtained by using some other energy source. Usually it is obtained by the electrolysis of water, or by breaking down natural gas (methane CH4). Hydrogen is highly explosive, and to be contained and carried in significantly usable amounts it has to be compressed or cooled to a liquid at minus 2530C. Hydrogen is not easy to handle, and it is not a convenient replacement for pouring 10 gallons of gasoline into an automobile fuel tank.

Fuel cells are being developed for use in transportation (automobiles, trucks, buses, etc.) but fuel cells have to be fueled with hydrogen. Fuel cells are not a source of energy in themselves, but are a possible ultimate substitute for the internal combustion engine. However, putting the infrastructure in place to effectively and economically produce and store hydrogen on the widespread basis as oil and its derivatives are today, is an enormous, costly, and long term task. The ultimate result can hardly be as versatile and convenient as is the use of oil products today around the world.

The limits of technology

We now live in very fortunate times. In the combination of the versatility of end uses, energy density, ease of handling and storage, and being now able to produce it relatively inexpensively and in great volume, there is no energy source comparable to oil. But living in a chiefly petroleum fueled economy and in a fossil fuel economy in general, we are living off our capital, which is unsustainable.

In a very perceptive volume for the time it was written [1952], British physicist [Charles] G. Darwin recounts the several "revolutions" which have taken place in the progress of human history, such as the most recent one, the Industrial Revolution. He states there is one more revolution coming:

> The fifth revolution will come when we have spent the stores of coal and oil that have been accumulating in the earth during hundreds of millions of years . . . it is obvious that there will be a very great difference in ways of life . . . a man has to alter his way of life considerably, when, after living for years on his capital, he suddenly finds he has to earn any money he wants to spend. . . . The change may justly be called a revolution, but it differs from all the preceding ones in that there is no likelihood of its leading to increase in population, but even perhaps to the reverse.

There is a popular belief that somehow technology can indefinitely rescue the human race from whatever predicament it may get itself into—solve all problems. D. Pimentel and M. Giampietro have warned:

> Technology cannot substitute for essential natural resources such as food, forests, land, water, energy, and biodiversity . . . we must be realistic as to what technology can and cannot do to help humans feed themselves and to provide other essential resources.

[Physics professor Albert A.] Bartlett (1994) has observed:

> There will always be popular and persuasive technological optimists who believe that population increases are good, and who believe that the human mind has unlimited capacity to find technological solutions to all problems of crowding, environmental destruction, and resource shortages. These technological optimists are usually not biological or physical scientists. Politicians and business people tend to be eager disciples of the technological optimists.

Growing world population

This is not to say that technology cannot continue to produce many good things in the future. But we must not confuse technology which uses resources with creating the resources. The world is finite; there are limits. Nature has given us a great inheritance formed in the Earth by myriad geological processes over millions of years consisting of a huge variety of resources, including, importantly now, fossil fuels. This is a nonrenewable bank account against which we have been writing larger and larger checks as the needs of an increasingly industrialized growing world population have been supplied.

But eventually this account will be exhausted, and we will have to bestir ourselves to get out and live on current income, the first need of which apparently will be to replace oil. How many people can a renewable energy resource income support? And what will be the resources we will use to do this?

[Population analyst Joel E.] Cohen has discussed this, as is the title of his book, "How Many People Can the Earth Support?" But, perhaps the question should be phrased "how many people should the Earth support?"

The optimum size of this population can hardly be estimated now with any great degree of accuracy. . . . It is significant that most of them determine a figure which is substantially smaller than is the size of today's population.

[Professor "Ted" F.E.] Trainer, in a comprehensive study of renewable energy sources, has made a well-supported clear statement:

> Figures commonly quoted on costs of generating energy from renewable sources can give the impression that it will be possible to switch to renewables as the foundation for the continuation of industrial societies with high material living standards. Although renewable energy must be the sole source in a sustainable society, major difficulties become evident when conversion, storage and supply for high latitudes are considered. It is concluded that renewable energy sources will not be able to sustain present rich world levels of energy use and that a sustainable world order must be based on acceptance of much lower per capita levels of energy use, much lower living standards and a zero growth economy.

Transition to an entirely renewable sustainable energy resource economy with resulting changes in lifestyles is inevitable. Will it be done with intelligence and foresight or will it be done by harsh natural forces? This is one of the main challenges which lie before us.

It seems likely that a sustainable energy mix will be broader than it is today where oil and natural gas make up more than 50% of our supplies. And energy in total will likely be more costly than our energy bill today. The transition to this wider diversity of energy sources will proceed slowly and probably be somewhat provincial depending on what regional resources are available.

Energy is the key which unlocks all other resources, and it will continue to be the key to human physical prosperity. It is significant that both the per capita use of oil, and the per capita use of energy in total both peaked in 1979 and have been falling ever since. We may already be seeing the beginning of the fifth revolution to which Darwin referred.

The British scientist and statesman, Sir Crispin Tickell has clearly summed up our situation:

> We have done remarkably little to reduce our dependence on a fuel [oil] which is a limited resource and for which there is no comprehensive substitute in prospect.

The challenge of conversion to alternative energy sources with the concurrent problems of population size and stabilization, and adjustment of economies and lifestyles is clearly at hand. A realistic appraisal of the future encourages people to properly prepare for the coming events. Delay in dealing with the issues will surely result in unpleasant surprises. Let us get on with the task of moving orderly into the post-petroleum paradigm.

4

Coal Power Harms the Environment

Sierra Club

The Sierra Club is an environmental organization that promotes conservation by influencing public policy decisions.

The United States relies primarily on old, dirty coal-powered generators for most of its electricity. Although these plants emit tons of toxic nitrogen oxide, sulfur dioxide, and mercury, the government is doing little to clean them up. In fact, federal regulators have made it easier for these plants to operate even as the lethal emissions from coal contribute to fatal lung disease, damaged ecosystems, and global warming.

We all use electricity in our daily lives, almost without thinking about it—turning on the lights, listening to the radio, and using computers. If we stopped and learned about the energy we use, we would encounter some shocking realities about the impacts of the energy production process on the environment and our health.

With all the amazing technological advancements over the last century, one thing that has not changed very much is our reliance on fossil fuels, in particular, dirty coal to generate electricity. More than half of the electricity generated in the United States comes from coal. As the producer of the largest share of our nation's energy, coal-fired plants are also some of the dirtiest.

Many older coal-fired power plants have enjoyed a loophole in the Clean Air Act, allowing them to avoid modernizing with pollution controls. As a result, as many as 600 existing power plants are between 30–50 years old and are up to 10 times dirtier than new power plants built today. When the Clean Air Act was proposed [in 1970] this loophole was included to get it passed because Congress assumed that newer plants would come into compliance with the Clean Air Act standards and soon replace the older more polluting plants. For a variety of reasons, including efforts to heavily subsidize coal, this has not happened. Therefore, we are now faced with a disproportionate amount of pollution coming from these old, dirty, under-controlled plants.

Out of the entire electric industry, coal-fired power plants contribute 96% of sulfur dioxide emissions, 93% of nitrogen oxide emissions, 88% of carbon dioxide emissions, and 99% of mercury emissions.

Smog

When nitrogen oxide (NO_x) reacts with volatile organic compounds (VOCs) [gasoline, industrial chemicals, and solvents] and sunlight, ground level ozone, or smog forms. Power plants are second only to automobiles as the greatest source of NO_x emissions. NO_x emissions from huge dirty coal plants with tall smokestacks in the midwest are often blamed for increased smog levels in many eastern regions because smog and its precursor pollutants are easily transported hundreds of miles downwind from pollution sources. More than 137 million Americans continue to breathe unhealthy, smog polluted air.

> *More than 137 million Americans continue to breathe unhealthy, smog polluted air.*

Even our national parks have not escaped the smog caused by coal-fired power plants. Regional haze from airborne pollutants has reduced annual average visibility in the US, to about one-third in the west and to one-quarter in the east, of natural conditions. Smog concentrations increased at 22 of 31 National Park Service monitoring sites from 1990–1999.

When inhaled, smog causes a burning of the cell wall of the lungs and air passages. This eventually weakens the elasticity of the lungs, making them more susceptible to infections and injury and causing asthma attacks and other respiratory illnesses.

This danger is present for anyone who inhales smog, although children, elderly, and those with respiratory problems are at a higher risk of developing health problems associated with smog pollution. A UCLA School of Medicine study found that over time, repeated exposure to smog and other air pollutants can cause as much damage to the lungs as smoking a pack of cigarettes a day. In addition, a recent . . . study found that high smog levels in the eastern US cause 159,000 trips to the emergency room, 53,000 hospital admissions, and 6 million asthma attacks each summer.

Soot

The burning of coal emits sulfur dioxide (SO_2) and nitrogen oxide (NO_x) gases, which can form fine particles, or soot, when they react with the atmosphere. In addition, coal-fired power plants also emit soot directly from their smokestacks. Scientists increasingly believe soot to be the most dangerous air pollutant, blaming it for 64,000 deaths per year in the US, which is almost twice the number of deaths due to auto crashes. Cutting power plant pollutants by 75% would avoid more than 18,000 of those deaths.

Soot causes bacterial and viral respiratory infections like pneumonia, as well as chronic lung diseases, like asthma, that destroy lives over the course of years. Soot from power plants triggers an estimated 603,000 asthma attacks nationwide every year. Bringing old plants up to modern standards would avoid 366,000 of these attacks. In addition, studies have found that soot may cause heart attacks and arrhythmia (irregular heartbeat) and that the incidence of strokes and heart failure is greater in areas with high levels of soot.

Acid rain

Acid rain is formed when sulfur dioxide (SO_2) and nitrogen oxide (NO_x) react with water and oxygen in the atmosphere to form acidic compounds, most commonly sulfuric and nitric acid. These compounds can become incorporated into natural precipitation and fall to the earth as rain or snow. Coal-fired power plants are the largest source of SO_2, 66%, and second to automobiles in NO_x emissions. The Northeast and eastern Canada are home to some of the worst acid rain pollution, because emissions produced from large dirty midwestern coal power plants waft in the wind toward the Northeast. For instance, numerous lakes and streams in the Adirondack mountains of upstate New York are too acidic to support fish life, and half of Virginia's native trout streams have reduced capacity due to acid rain.

> *Numerous lakes and streams in . . . upstate New York are too acidic to support fish life, and half of Virginia's native trout streams have reduced capacity due to acid rain.*

Acid rain destroys the ecosystems, including streams and lakes, upsetting the delicate balance and making them unable to support life. It also can destroy forests, killing plant and animal life and eats away at man-made monuments and buildings, effectively destroying our natural and historical treasures.

While the 1990 amendments to the Clean Air Act have made great progress in reducing SO_2 emissions from many of the midwestern coal power plants, more needs to be done. Too many of the lakes and streams in our country continue to suffer from the devastating effects of acid rain.

Toxins

Power plants are one of the largest sources of toxic metal compound pollution. Together they released more than one billion pounds of toxic pollution in 1998, including 9 million pounds of toxic metals and metal compounds and 750 million pounds of dangerous acid gases. Many of these compounds are known or suspected carcinogens and neurotoxins and can cause acute respiratory problems, and aggravate asthma and emphysema.

One of the most dangerous toxins emitted is mercury. Coal contains trace amounts of mercury that are released into the air when the fuel is burned to produce electricity. The health hazard results when mercury falls to the earth with rain, snow, and in dry particles.

Mercury is a serious toxin, and accidental high-level exposure can result in severe nervous system damage, even death. But exposure to toxic mercury primarily affects fetal development. In unborn children, it can influence the development of the brain and nervous system. When infants are exposed to toxic mercury by their mothers through breast milk, the result can be extremely dangerous and can cause delays in walking, talking, and fine motor skills. The primary exposure pathway for most Americans is through consumption of fish with high levels of methyl mercury, the toxic form of mercury that accumulates in fish and shellfish and the animals that eat those fish, including humans. More than 70% of the fish advisories issued in 2002 were for mercury contamination.

Global warming

Burning fossil fuels such as coal releases carbon dioxide (CO_2) pollution. The US has four percent of the world's population yet emits 25% of the global warming pollution. Power plants emit 40% of US carbon dioxide pollution, the primary global warming pollutant. In 1999, coal-fired power plants alone released 490.5 million metric tons of CO_2 into the atmosphere (32% of the total CO_2 emissions for 1999). Currently there is 30% more CO_2 in the atmosphere than there was at the start of the Industrial Revolution [in the late eighteenth century], and we are well on the way to doubling CO_2 levels in the atmosphere during this century.

The 1990s were the hottest decade on record. Average global temperatures rose one degree Fahrenheit during the last century and the latest projections are for an average temperature increase of two to as much as ten degrees during this century. In February 2001, the Intergovernmental Panel on Climate Change (IPCC) reported that global warming threatens human populations and the world's ecosystems with worsening heat waves, floods, drought, extreme weather and by spreading infectious diseases. To address the problem of global warming, steps need to be taken to slash the amount of CO_2 power plants emit. We need to switch from burning coal to cleaner burning natural gas and dramatically increase energy efficiency and renewable wind and solar energy.

The government should expand the Clean Air Act to include protections from old and dirty power plants and provide incentives for the use of cleaner fuels. The government should also work towards the replacement of the existing infrastructure with a more sustainable means of producing electricity.

Individuals can help by conserving electricity in the home and office.

5

Nuclear Power Is Efficient and Safe for the Environment

Nuclear Energy Institute

The Nuclear Energy Institute (NEI) is the policy organization of the nuclear energy industry whose objective is to promote policies that benefit the nuclear energy business.

Nuclear energy is a clean, stable way to generate power. Unlike coal, gas, and oil generators, nuclear power plants do not emit toxic pollutants into the air. Nuclear plants are relatively small and have less of a "footprint" on the land than wind or solar farms. The areas around nuclear plants are often parklike habitats that are home to many types of endangered species. Compared to many sources of electrical generation, nuclear power plants are relatively benign.

The use of nuclear energy has increased in the United States since 1973. Nuclear energy's share of U.S. electricity generation has grown from 4 percent in 1973 to almost 20 percent in 1999. Part of the increase is due to improved plant performance. Just since 1990, the increased output from the nation's nuclear plants has been the equivalent of bringing 19 new 1,000-megawatt nuclear plants on line.

This is excellent news for the environment. Nuclear energy and hydropower are the two large-scale means of producing electricity while keeping the air clean. Because nuclear power plants do not burn fuel, they emit no combustion byproducts—like air pollutants and carbon dioxide—into the atmosphere.

Emissions of nitrogen oxide and sulfur dioxide are regulated by the 1990 Clean Air Act amendments. . . .

Nitrogen oxide (NO_x) plays a major role in the formation of ozone, which is detrimental to human health. NO_x is also a significant contributor to acid rain. . . .

By substituting for fossil fuels in electricity generation, U.S. nuclear power plants currently avoid almost two million tons of NO_x emissions annually. . . . Between 1973 and 1999, nuclear energy avoided emission of 31.6 million tons of NO_x. . . .

Sulfur dioxide (SO_2) is thought to contribute to acid rain. A main objective of the Clean Air Act amendments is to reduce the amount of SO_2 emitted into the atmosphere. Between 1990 and 1995, generation from nuclear power plants serving the states affected by the act's initial emission reduction targets increased by more than 16 percent. By displacing fossil fuels to generate electricity, this increased generation avoided 480,000 tons of (SO_2) emissions. . . .

> *Because nuclear power plants do not burn fuel, they emit no combustion byproducts—like air pollutants and carbon dioxide—into the atmosphere.*

Since the 1973 oil embargo nuclear energy has contributed even more significantly to U.S. air quality. By substituting for fossil fuels, U.S. nuclear power plants displaced a cumulative total of 61.9 million tons of (SO_2) between 1973 and 1999.

Reducing carbon dioxide emissions

As sunlight passes through the air and reaches the ground, it turns into heat. Certain gases in the atmosphere act like the glass in a greenhouse, preventing some of this heat from escaping back into space. This trapped heat helps keep the Earth comfortably warm.

But many scientists believe that carbon dioxide emissions from human activities add to the warming effect, bringing about changes in climate. . . .

Carbon dioxide is estimated to be responsible for one-half of any global warming.

By substituting for fossil fuels, U.S. nuclear plants reduced total U.S. greenhouse gas emissions by 168 million metric tons of carbon equivalent in 1999. Without nuclear energy, U.S. electric utility emissions of carbon equivalents would have been approximately 30 percent higher.

> *Radiation levels at every plant are monitored 24 hours a day, seven days a week.*

Generating one million kilowatt-hours of electricity produces about 150 metric tons of carbon from a natural gas–fired plant, 265 metric tons from a coal-fired plant and 220 metric tons of carbon from an oil-fired plant—but no carbon from a nuclear power plant. (In the United States, coal-fired power plants supply electricity to the facilities that enrich uranium for fuel. About 10 metric tons of carbon are emitted from these plants in the enrichment of enough fuel to produce one million kilowatt-hours of electricity.)

Long term, nuclear energy reduced total U.S. CO_2 emissions by 2.61 billion metric tons of carbon between 1973 and 1999, by replacing fossil fuels for electricity generation.

Worldwide, nuclear energy has significantly reduced greenhouse gas emissions. Approximately 430 nuclear power plants in 31 nations produce 17 percent of the world's electricity—while reducing CO_2 emissions by some 500 million metric tons of carbon. . . .

Strict standards, careful control

All methods of producing electricity affect the environment to some degree, but the impacts from nuclear energy are minimal—one of the lowest on a per-kilowatt-hour basis.

Because the fuel in nuclear power plants is radioactive, nuclear plants are carefully designed, built and monitored to prevent releases of radioactive material. The Environmental Protection Agency [EPA] sets—and the NRC [Nuclear Regulatory Commission] enforces—strict standards governing radiation emissions.

To make sure that nuclear power plants operate well within those standards, radiation levels at every plant are monitored

24 hours a day, seven days a week. Even soil, cows' milk from neighboring farms, and fish and sediment in nearby rivers and lakes are monitored periodically. The monitoring instruments are so sensitive that they can measure even trace amounts of radiation. Nuclear power plant emissions are always well below the safe levels permitted by federal standards. That is why the environment has never been harmed by radiation emissions from a U.S. nuclear power plant.

Even the people living closest to a nuclear power plant receive an average of only one extra day's worth of radiation—about one millirem—each year. In comparison, the average American is exposed to 360 millirem annually from the natural environment and man-made sources, like medical X-rays.

Protecting aquatic resources

Before a plant begins operating, an environmental impact statement examines all potential impacts to water quality from the operation of the plant. These include concerns about the discharge of heated water and the possibility of trapping aquatic life in the intake. All issues are resolved by the time the plant is licensed. If a license is later renewed, the plant must certify that no significant adverse impacts have been observed during the plant's operating life.

Like all steam-electric generating plants, nuclear power plants must take in water for cooling. That is why many of them are located on rivers, lakes and bays. After it is used for cooling, the water—now slightly warmed—needs to be discharged. (This water has never come in contact with radioactive materials.)

> *Electric utilities voluntarily work to protect the fish, mammals, reptiles, birds and plants found on or near power plant sites.*

Cooling water discharged from a plant contains no harmful pollutants, but still must meet federal Clean Water Act requirements and state standards designed to protect water quality and aquatic life. If the water is warm enough to possibly harm aquatic life, it is cooled before it is returned to its source river, lake or bay. It is either mixed with water in a cooling pond or pumped through a cooling tower before it is discharged. In addition, power plants operate under National Pollutant Discharge Elimination System permits, which specify standards and monitoring requirements for all water discharges from the plants. These permits, which must be renewed every five years, require plants to use the best technology available, thus minimizing environmental impacts.

The Nuclear Regulatory Commission also reviews plant operations to be sure there is no adverse impact to water quality and aquatic ecology. Many early aquatic resource concerns have not materialized at any nuclear power plant.

Protecting wildlife and habitats

Because the area around a nuclear power plant is so clean, the areas around cooling ponds are often developed as environmentally rich wetlands, providing better nesting areas for waterfowl and other birds, new habitats for fish, and preservation of other wildlife, flowers and grasses.

Electric utilities voluntarily work to protect the fish, mammals, reptiles, birds and plants found on or near power plant sites. Many have created special nature parks or wildlife sanctuaries on plant sites.

For example, Virginia Power protects a bald eagle nesting site at its Surry nuclear plant and nesting boxes for wood ducks and barn swallows at its North Anna nuclear plant. It also built 20 underwater block-and-brush structures in Lake Anna, where young fish can find cover and large fish can feed and spawn. When the Turkey Point nuclear power plant in Florida dredged some 160 miles of cooling canals, they became a safe nesting ground where newly hatched crocodiles—often hunted for their skin—have a chance to survive.

Managing spent fuel

Management of used nuclear fuel is one of the most successful solid waste management programs ever for dealing with the byproduct material of our industrial society. The fuel is radioactive and, therefore, is kept safely stored away from the environment.

What is used fuel? Like other power plants, nuclear plants create electricity by boiling water into steam, which turns a turbine-generator. Nuclear power plants do not burn anything to create this heat. Instead, they fission—or split—uranium atoms in a chain reaction. This is a clean, non-polluting process.

Uranium fuel, in the form of small ceramic pellets, is placed inside metal fuel rods, which are grouped into bundles, called assemblies. Over time, the fuel's energy is consumed. Thus, every 18–24 months the reactor is shut down and the oldest fuel assemblies—which have released their energy but have become radioactive as a result of fission—are removed and replaced.

Relatively small volume—All of the country's nuclear power plants together produce about 2,000 metric tons of used fuel annually. All the used fuel ever produced by the U.S. nuclear energy industry in more than 40 years of operation—some 40,000 metric tons—would cover an area only the size of a football field to a depth of about five yards, if the fuel assemblies were stacked side by side and laid end to end.

Losing its radioactivity—Used fuel is highly radioactive when it is removed from the reactor, but it loses its radioactivity as time goes by. Most used fuel loses about 50 percent of its radioactivity after three months and about 80 percent after one year. Less than 1 percent will remain radioactive for thousands of years. (In contrast, chemical waste remains toxic forever.) All used fuel is carefully isolated from people and the environment.

Safe storage Today, this used fuel is stored at the plant sites, either in steel-lined, concrete vaults filled with water, called used fuel pools, or in above-ground steel or steel-reinforced concrete containers with steel inner canisters. On-site storage is an interim measure, however, and licenses issued by the NRC limit the amount of used fuel that a utility is permitted to keep on site. Although the NRC determined that used fuel could be stored at plant sites for 100 years without adverse health or safety consequences, it also believes that timely disposal is necessary.

In the Nuclear Waste Policy Act of 1982 and its 1987 amendments, Congress created a timetable for a long-term solution: a deep, mined geologic repository built in an unpopulated desert area in Nevada. A scientific study of that site, called Yucca Mountain, is under way and is nearing determination of its suitability.

6

Nuclear Power Is Inefficient and Dangerous

Amory B. Lovins and L. Hunter Lovins

Amory B. Lovins and L. Hunter Lovins are co-CEOs of the Rocky Mountain Institute, a think tank that emphasizes market-based solutions on economic and environmental issues. Both are longtime advisers to the energy industry and the U.S. Departments of Energy and Defense. Together they have coauthored twenty-seven books.

The nuclear power industry has a dismal record in terms of power costs, disposal of radioactive waste, and plant safety. Once billed as the cure to the world's energy problems, it has become obvious over the decades that without massive government support, the nuclear energy industry would be bankrupt. The enriched uranium used to generate power in nuclear plants is one of the most toxic substances on earth and remains a deadly poison for tens of thousands of years. With new developments in hydrogen, solar, and wind power technologies, the environmental and economic costs incurred by nuclear power plants cannot be justified.

B uoyed by a supportive [Bush] White House, growing climate concerns, temporarily high gas prices, and California's [2001] electricity mess, the nuclear industry is running an all-out public-relations campaign to resuscitate its product. This attempt ignores one crucial fact: Nuclear power already

Amory B. Lovins and L. Hunter Lovins, "Fuel Cell R&D Is Far from Easy Street," *Electronic Engineering Times*, May 26, 2003. Copyright © 2004 by CMP Media LLC, 600 Community Dr., Manhasset, NY 11030, USA. Reproduced by permission.

died of an incurable attack of market forces. Once touted as "too cheap to meter," nuclear power, as *The Economist* recently concluded, now looks "too costly to matter."

Overwhelmed by huge construction and repair costs around the world, nuclear plants ended up achieving less than 10% of the capacity and 1% of the new orders (all from countries with centrally planned energy systems) forecast a quarter-century ago. The industry has suffered the greatest collapse of any enterprise in industrial history. Beyond the hard economic facts, about which more later, the nuclear industry is dismissing legitimate public concerns about the risks of a technology so unforgiving that, as Nobel physicist Hannes Alfven wrote, "No acts of God can be permitted." Each nuclear plant, through accident or malice, could release enough radioactivity to hazard a continent. This is presented by the industry as extremely unlikely, but many citizens aren't reassured. They have seen too many highly improbable events, including terrorism. And if nuclear power plants are so safe, why would the industry build and run them only if the federal government passed a law limiting operators' liability in major accidents? Why should the nuclear industry enjoy a liability cap that reduces its incentive for safety, distorts choices with a vast subsidy and is unavailable to any other industry? Why can't nuclear operators self-insure and put their money where their mouths are, or buy insurance at market prices like everyone else? . . .

> *Each nuclear plant, through accident or malice, could release enough radioactivity to hazard a continent.*

Scientists still haven't developed reliable ways to handle nuclear wastes and decommissioned plants, which remain dangerously radioactive for far longer than societies last or geological foresight extends. And experts feel nuclear power's gravest risk is that power plants can provide ingredients and innocent-seeming civilian cover for the development of nuclear bombs, as was the case in India and elsewhere. Now the White House proposes to revive nuclear-fuel reprocessing after decades of proof that it's unprofitable, unnecessary, a complication to nuclear waste management and a source of vast amounts of bomb material.

Uncompetitive and unnecessary

Market economics provides an even more basic argument: "If a thing is not worth doing," said economist John Maynard Keynes, "it is not worth doing well." Leaving aside bomb-proliferation, waste, sabotage and uninsurable accidents, nuclear power is simply uncompetitive and unnecessary. After a trillion-dollar taxpayer investment, it delivers little more energy in the U.S. than wood. Globally, it produces severalfold less energy than renewable sources. The market prefers other options. In the 1990s, global nuclear capacity rose by 1% a year, compared with 17% for solar cells (24% in 2000) and 24% for wind power—which has lately added about 5,000 megawatts a year worldwide, as compared with the 3,100 new megawatts nuclear power averaged annually in the 1990s. The decentralized generators California added in the 1990s have more capacity than its two giant nuclear plants—whose debts triggered the restructuring that created the state's current utility mess.

> *The nuclear industry has a well-earned reputation for breezy mendacity.*

Enthusiasts claim new-style reactors might deliver a kilowatt-hour to your meter for 5 cents, compared with 10 to 15 cents for post-1980 nuclear plants worldwide. (Of that, 10 to 15 cents, nearly 3 cents pays for delivery, about 2 cents for running the plant, and the rest for its construction and for occasional major repairs.) But on the same accounting basis, superefficient gas plants or wind farms cost only 5 to 6 cents per kilowatt-hour, cogeneration of heat and power often 1 to 5 cents, and efficient lights, motors and other electricity-saving devices under 2 cents, often under 1 cent. Cogeneration and efficiency are especially cheap because they occur at the site where the energy is consumed and thus require no delivery.

All these non-nuclear options continue to get cheaper, as do fuel cells and solar cells. Today, a pound of silicon [in a solar cell] can produce more electricity than a pound of nuclear fuel. Already, Sacramento's municipal utility, which has successfully replaced power from its ailing nuclear plant (shut down by voters) with a portfolio emphasizing efficiency and renewables, has brought the heretofore costliest option, solar cells, down to costs competitive with a new nuclear plant.

Efficiency saves power

The PR spinners trumpet that nuclear power costs less than power from gas plants. This is true if you are looking only at the running cost of an average existing nuclear plant, compared with the running costs of an old, inefficient gas-fired plant. It does not include delivery to customers, nor the prohibitive construction costs of a new nuclear plant. Notice, too, the ads don't compare the costs of a new nuclear plant with the new, doubled-efficiency gas plants that are beating the pants off nuclear and coal worldwide. Under such realistic cost comparisons, nuclear power plummets to its actual status as the worst buy available. You didn't understand that from those slick ads? You weren't supposed to. The nuclear industry has a well-earned reputation for breezy mendacity.

Lost in the debate over what kind of new plant to build is the best option of all: more efficient use of the electricity we already have. We've been reducing electricity use per dollar of gross domestic product by 1.6% a year nationwide, and in California between 1997 and 2000, by 4.4% a year. California has held its per-capita electricity use essentially flat since the mid-1970s, yet far more savings remain untapped—enough nationally to save four times nuclear power's output, at one-sixth its operating cost. Our personal household electric bill, for example, is $5 a month for a 4,000-square-foot house in the Rocky Mountains. Passive solar design and super-efficient appliances and lighting yielded a 90% savings on electricity and 99% on fuel. The improvements, made in 1983, paid for themselves in 10 months. Today's technologies are far better. An estimated three-fourths of U.S. electricity could now be saved through efficiency techniques that cost less than generating that power, even in existing plants. . . .

After a half-century of nuclear power, the verdict of the marketplace is in. Nuclear power has flunked the market test. Nuclear salesmen scour the world for a single order, while makers of alternatives enjoy brisk business. Let's profit from their experience. Taking markets seriously, not propping up failed technologies at public expense, offers a stable climate, a prosperous economy, and a cleaner and more peaceful world.

7

America's Energy Security Depends on Developing Energy Alternatives

Union of Concerned Scientists

The Union of Concerned Scientists is an independent non-profit alliance of scientists and citizens that seeks to use scientific research to solve social and environmental problems.

The United States relies heavily on oil, natural gas, coal, and nuclear energy. In addition to creating unacceptable levels of pollution, these energy sources have many inherent security problems that could create havoc in the lives of millions of people. To obtain enough oil, the United States must rely on foreign countries in which dictators or unstable governments can create oil shortages with little warning. In addition, gas pipelines and nuclear power plants may be attacked by terrorists with possible disastrous consequences for millions of people. The United States must move toward alternative energy sources such as solar, hydrogen, and wind in order to make America safer, cleaner, and less vulnerable to terrorist attacks.

As the world's largest oil consumer, the United States is particularly exposed to the risks posed by an oil market beyond our control. Reliance on the economically powerful

Union of Concerned Scientists, *Energy Security: Solutions to Protect America's Power Supply and Reduce Oil Dependence*. Cambridge, MA: Union of Concerned Scientists, January 2002.

OPEC cartel[1] and the politically unstable Persian Gulf nations will only grow over time as oil supplies dwindle. OPEC owns four-fifths of the world's remaining proven oil reserves and nations in the Persian Gulf own two-thirds. Only a small proportion—about 2 percent—of the proven reserves lies within the United States.

Importing large amounts of oil carries significant economic costs: we send more than $200,000 overseas each minute to buy foreign oil. But even if we imported no oil at all, the US economy would still be vulnerable. The price we pay for oil is determined by the world market, so global price hikes affect the cost of US oil because all oil retailers (domestic and foreign) charge more. As long as the US economy is tied to oil—and oil is traded globally—we will be susceptible to OPEC's market power and Persian Gulf instability. To date, the economic costs of oil dependence have been tremendous, totaling $7 trillion over the past 30 years by one estimate.

The political instability of the Persian Gulf has caused three major price shocks over the past 30 years. The Iraqi invasion of Kuwait in 1990 took an estimated 4.6 million barrels per day out of the global oil supply for three months. The Iranian revolution reduced global oil supplies by 3.5 million barrels per day for six months in 1979, and the Arab oil embargo eliminated 2.6 million barrels per day for six months in 1973. In each of these cases, the world oil supply dropped only about 5 percent, but world oil prices doubled or tripled.

> *We send more than $200,000 overseas each minute to buy foreign oil.*

Dramatic changes in the US economy have closely tracked world oil prices. In the wake of oil price hikes, US inflation increased markedly, accompanied by downturns in our gross domestic product. In each case, recession followed. . . .

Not only does reliance on oil from the unstable Persian Gulf increase our economic risks, it drives our foreign policy in

---

1. OPEC, the Organization of Petroleum Exporting Countries, consists of Algeria, Gabon, Indonesia, Iran, Iraq, Kuwait, Libya, Nigeria, Qatar, Saudi Arabia, the United Arab Emirates, and Venezuela.

the region. Direct military expenditures associated with defending Persian Gulf oil supplies have been estimated at $20 to $40 billion per year. In addition, US oil purchases are funding Persian Gulf countries to the tune of $28 billion per year, including $18 billion per year to Saudi Arabia and $7 billion per year to Iraq. Our support for totalitarian states with dismal human rights records and links to terrorism place a heavy burden on US security and foreign policy.

Security risks

Much of the US energy system presents significant safety and security risks. Facilities recently put on heightened security alerts include nuclear power plants, hydropower dams, pipelines, refineries, tankers, and the electricity transmission grid. The cause for concern is well placed: past disruptions—whether through sabotage, natural disaster, or equipment failure—have affected our economy and, in some cases, public safety. Electricity outages alone cost the United States tens of billions each year.

Our reliance on vulnerable infrastructure will escalate as our demand for energy grows. For example, the Bush administration's National Energy Policy—which focuses on increasing energy supply rather than conservation—is based on a forecast need for 1,300 to 1,900 new power plants over the next 20 years. Since most of these plants will use natural gas to produce electricity, we will need to increase our supply of gas by nearly 25 percent by 2010 and add 301,000 miles of new natural gas transmission and distribution pipelines.

US energy production is highly centralized, making it vulnerable to well-placed attacks. A typical nuclear power plant serves 1.25 million homes, a large coal power plant half a million homes. Concentration in the oil business is even greater, with only 150 refineries nationwide producing fuel for over 200 million cars. A major disruption at any one of these energy supply facilities can endanger lives and affect prices for hundreds of thousands of customers.

Because nuclear power plants rely on radioactive materials and create radioactive waste, they present unique safety risks. Government studies report that the radioactive material released through damage to either the reactor or the spent fuel stored on site could kill or injure tens of thousands of people living within 500 miles of the plant and render an area the size of Pennsylvania uninhabitable for generations. . . .

The US network for distributing fuel and electricity is also highly vulnerable. Accidents or sabotage at the key hubs of pipelines, transmission lines, and tanker/barge systems could disrupt energy flow to large portions of the country.

Large oil pipelines that serve as the central artery for fuel are particularly vulnerable. For example, the Trans-Alaska Pipeline System (TAPS) stretches across 800 miles, about half of the distance above ground. Each day it carries about 1 million barrels of oil across the state. Beginning October 4, 2001, TAPS was closed for three days after a hunter shot a hole in the pipeline. In addition to disrupting supply, such events carry environmental impacts and economic costs via property damage. In 2000, 147 pipeline incidents reported in the United States spilled over 100,000 barrels and caused more than $100 million in property damage. . . .

Power failures

The system that delivers our electric power is technically complex and highly vulnerable. The nation's electricity grid includes over 150,000 miles of mostly unprotected high-voltage transmission lines, control centers, substations, and transformers, often in physically remote locations. Because electricity cannot be easily stored, an interruption of the grid results in instantaneous power loss. Twice in 1996, millions of customers across the Western United States lost power when the electric power grid failed. What's more, a sudden loss of a major component of the transmission grid, if not responded to quickly, can result in cascading blackouts.

> *A major blackout would have far-reaching effects including disabling traffic lights, disrupting air traffic control, and shutting down the systems that protect our water supplies.*

Some parts of the grid system are critical to maintaining power to large geographic areas. For example, a single substation is the connection point both for the transmission lines between southern and northern California and for power plants that provide energy to the San Francisco area. If one of these essential components of the system were disabled, it could take several weeks to restore power. A major blackout would have far-reaching effects including disabling traffic lights, disrupting air traffic control, and shutting down the systems that protect our water supplies. Blackouts over a wide geographic area or a long period could wreak economic damage as well.

Like most other systems that make up our national infrastructure, the electricity grid depends on computer information networks and telecommunication systems connected through the Internet. In 1997, a federal agency [the President's National Security Telecommunications Advisory Committee] concluded that an adequately funded terrorist group "could conduct a structured attack on the electric power grid electronically without having to set foot in the target nation and with a large degree of anonymity." This was borne out in the spring of 2001 when Internet hackers invaded the computers that control California's power system for over two weeks. While the breach did no apparent harm, the investigation detected signs that the hackers were planning an attack on the system.

Global warming

America's current energy system, with its oil dependence and infrastructure vulnerability, poses clear and present dangers. But our energy use also contributes to global warming, which presents additional, and profound, long-term risks to our economic, environmental, and national security. Climate change is likely to lead to rising sea level, more extreme storms, more erosion, flooding, droughts, and heat waves, poorer air quality, greater risk of disease, and disruption of agriculture.

The risks to the US energy infrastructure, such as more extreme storms or flooding, are clear. But the indirect security risks could be more significant. The costs of climate change are unlikely to be borne equally among the world's peoples, as poorer countries may not have the financial resilience to cope with the impacts. If unabated, the increasing rift between rich nations such as the United States and poorer, disaffected peoples could grow, leading to greater anti-American sentiment abroad. As the world's largest emitter of the heat-trapping gases that cause global warming, the United States carries a responsibility to address the threats posed by climate change, both as a global citizen and for its own security.

Secure energy solutions

We cannot change our energy system overnight, but we can—and must—make investments today that will create a more secure energy future tomorrow. Efficient cars, homes, and industry will deliver vital oil savings within the decade. Renewable power and efficient appliances will eliminate the need for hundreds of new large power plants in the coming years. Together, these solutions will produce an energy future that is far safer, cheaper, and cleaner than our current energy system.

Many of the risks posed by America's energy system are not captured in the prices we pay for energy. For example, we do not pay the military costs of protecting Persian Gulf oil at the gasoline pump. We do not pay to avoid global warming's impacts in our monthly electricity bills. But these are real costs that we as a nation will pay, they will just be hidden in military budgets or emergency relief expenditures or increased health costs. Incorporating the costs of security and environmental impacts should be a priority and doing so will spur us to adopt the solutions described below.

Clean, efficient, and safe vehicles

Passenger vehicles are the key to our oil dependence. Cars and light trucks currently account for 40 percent of all petroleum demand, but that fraction will approach half of all oil use by 2020, if recent trends continue. A leading cause of rising oil use is the fact that the fuel economy of new vehicles is the lowest it has been since 1980. Fuel economy standards set in 1975, the so-called Corporate Average Fuel Economy (CAFE) standards, effectively doubled the fuel economy of the new vehicle fleet over the following 10 years. But they have remained virtually untouched since 1985.

While CAFE standards remain stagnant, the vehicle market has undergone a dramatic transformation. Today, more than half of all new vehicles sold are light trucks, a category that includes SUVs, minivans, and pickups. As a result, fuel use has jumped. Under CAFE laws, light trucks must meet an average fuel economy standard of only 20.7 miles per gallon (mpg), while cars meet an average of 27.4 mpg. This loophole has created a major oil leak for the US economy. We burn 1.2 million more barrels of gasoline per day because our light trucks are less efficient than cars.

Fortunately, off-the-shelf technologies can affordably and safely boost the fuel economy of light trucks and cars. Studies from the Union of Concerned Scientists and the National Academy of Sciences both find that fuel economy could be dramatically improved over the next decade.

A fleet that relies on continuously evolving conventional technologies could reach an average of more than 40 mpg with improvements such as direct fuel injection, variable valve control engines, six-speed or continuously variable transmissions, high-strength lightweight materials, and low rolling resistance tires. These improvements would yield fuel savings of $3,000 to $5,000 over the lifetime of a car or truck, more than enough to make up for the cost of the changes. . . .

> *The United States carries a responsibility to address the threats posed by climate change, both as a global citizen and for its own security.*

Emerging technologies promise even greater oil savings, through both efficiency gains and a move to alternative fuels. Hybrid electric vehicles combine the electric motor and energy-storage system of an electric vehicle with the engine of a conventional vehicle. Toyota and Honda have successfully introduced hybrids into the American car market, and every major automaker is expected to introduce one in the next few years. Relying on hybrid electric vehicle technologies could increase fleet fuel efficiency to at least 55 mpg by 2020. These vehicles will save consumers $3,500 to $6,500 at the pump.

While hybrids offer a major leap forward in efficiency, fuel cells are often described as the technology to bring the century-long reign of the internal combustion engine to an end. Fuel cells combine hydrogen and oxygen to produce electricity, which then runs an electric motor. Current expectations are that family-sized fuel cell cars could reach a fuel economy of 80 mpg—triple that of today's vehicles. What is more, fuel cells run best on hydrogen, a nonpetroleum fuel which is made from natural gas today, but which can be manufactured from renewable sources in the future. Automakers are investing billions of dollars in the race to introduce a commercially viable fuel cell vehicle, a race likely to be won [by 2010]. . . .

Clean, safe power and energy efficiency

The nation's electricity generation is dominated by large, dirty, and unsafe power plants. Roughly half of our power is generated from coal and a fifth by nuclear plants. Electricity generation using natural gas is forecast to triple in the next 20 years, reaching 36 percent of the total, but recent price spikes are a reminder that this path carries severe economic risks.

Fortunately, clean renewable energy sources—such as wind, biomass, geothermal, solar, and landfill gas—can reduce our growing reliance on fossil fuel and nuclear generation, which depend on dangerous, polluting fuels and vulnerable delivery systems. Efficiency improvements in homes, businesses, and factories can further reduce our reliance on these unstable electricity sources. Furthermore, residential and industrial efficiency improvements can save oil, as these two sectors together account for 28 percent of the nation's oil demand.

Nuclear Energy. The future of nuclear power in the United States is uncertain due to safety hazards and unfavorable economics. Today, we are faced with immediate security risks from the 103 plants currently in operation. Future generations will live with risks from spent nuclear fuel, which will not diminish for thousands of years.

> *Using every last drop of oil from proven US reserves would fuel our economy for less than three years.*

The proposed new generation of nuclear power plants are no better. They would generate as much spent fuel as today's reactors. Even worse, the proposed design reduces costs by weakening the containment barriers installed to protect the public from accidents and sabotage.

Renewable Energy. Renewable energy sources carry none of the risks of nuclear power or even of fossil fuel generators. Electricity generated from wind, biomass, geothermal, solar, and landfill gas produces no hazardous wastes, is geographically dispersed, and requires less of the infrastructure that makes our current system so vulnerable.

Renewable energy technologies have made tremendous

strides in improving productivity and reducing costs over the years. In two decades, the cost for wind and solar electricity has come down by a factor of 5 to 10. Wind power and geothermal facilities at the best sites produce electricity for 3 to 6 cents per kilowatt-hour (¢/kWh), making them competitive with other electricity sources. But even when they cost more, renewable technologies help stabilize electricity prices: because their operating costs are low and their fuel is free, they create competitive pressure, which acts to restrain the price of fossil fuels like natural gas.

Combined Heat and Power. The efficiency with which fossil fuels are used can be greatly enhanced through dual generation of electricity and heat in a combined heat and power (CHP) facility. Where a conventional electricity generator might average 33 percent efficiency, some CHP technologies can reach efficiencies of 80 percent or better. And they are less vulnerable to disruption than are major power plants, because they are smaller and more dispersed. . . .

Consumer savings and environmental benefits

Not only do efficiency and renewable energy offer a secure strategy for cutting our reliance on oil and reducing the vulnerability of our energy infrastructure, they save money and protect the environment. Lower energy bills more than offset the incremental costs associated with renewable and efficiency technologies. Net savings will total $640 billion over the next two decades, with annual savings to consumers reaching $150 billion per year in 2020 or $500 for a typical family.

Renewable power and electrical efficiency can reduce carbon emissions from power plants by two-thirds from business as usual by 2020. Vehicle efficiency can deliver a 40 percent reduction in carbon emissions from passenger vehicles by 2020. These reductions will provide vital protection from the economic, environmental, and security threats of global warming. Similarly, clean power and vehicles can deliver large reductions in the pollutants that cause harmful smog, soot, and acid rain. For example, the technologies and policies we outlined here could reduce emissions of smog-forming pollutants from power plants and refineries by 3.3 million tons in 2020. In the power sector alone, emissions of sulfur dioxide and nitrogen oxides would decrease by more than 55 percent in that year.

False security

Despite the major security benefits offered by clean, efficient power and vehicles, some continue to push energy policies conceived before [the terrorist attacks on] September 11 [2001] that rely on vulnerable infrastructure and increasing domestic oil production in some of our last wild places.

The United States, with only 6 percent of the world's population, accounts for 26 percent of the world's oil demand. Because we possess only 2 percent of world oil reserves, however, anything we do to increase domestic production will have little or no effect on the global oil supply. We simply do not have enough oil left as a country to drill our way out of our predicament: using every last drop of oil from proven US reserves would fuel our economy for less than three years.

The only sensible strategy to reduce our oil dependence is to curb our appetite for petroleum. Clean and efficient vehicles, combined with homes and industries that are more efficient, can save more oil in just 10 years than the Arctic National Wildlife Refuge could economically produce in 60. Even at its peak (two decades from now), the Arctic Refuge would be producing 14 times less oil than can be saved through efficiency.

Continued reliance on nuclear power and fossil energy will only increase the vulnerability of our domestic infrastructure. New power plants, transmission lines, tankers, and pipelines will expand the hazards posed by our current energy system. Without strong policies to enhance security at nuclear power plants, curb our growing power demand, and shift us toward renewable energy, we are destined to face greater risks in the future.

8

The Use of Solar Energy Should Be Increased

Glenn Hamer

Glenn Hamer is executive director of the Solar Energy Industries Association in Washington, D.C.

In the future, a larger percentage of the worlds' energy needs will be taken care of by solar devices that harness the heat and light provided by the sun. Solar energy can be used passively to generate heat or be converted directly to electricity by photovoltaic cells. Although solar collectors have been prohibitively expensive in the past, prices are falling. Americans should increase their reliance on this efficient, environmentally friendly energy source.

The forecast for solar energy in the twenty-first century is sunny. Solar will push the trend of safe, affordable, and reliable distributed energy. As we head deeper into the new century, some predict that the electricity system may come to resemble the Internet—with disparate points and no one center of activity. In such a system, photovoltaics integrated into building roofs and windows will be a key component. Especially in the Southwest, concentrating solar power (CSP) stations will provide a significant percentage of the energy.

The solar photovoltaics industry grew at an annual rate of over 20 percent worldwide in the past decade and 40 percent [between 2000–2002], reaching an annual photovoltaic module production of 100 megawatts (MW) in the United States and about 400 MW globally [a typical coal power plant produces 500 MW]. Worldwide in 2001, photovoltaics was a $2 billion business. In the United States alone, the industry employs about 20,000 people in high-value, high-tech jobs.

In the swimming pool sector of solar thermal, according to the Florida Solar Energy Center, the equivalent of 594 MW of power was installed in 2001 in the United States. For solar hot-water heating for general home use, one utility, Hawaii Electric, has installed systems that produce approximately 60 MW per year, and it continues to add around 12 MW of capacity annually.

Solar energy technologies are of three primary types: photovoltaics (PV), concentrating solar power (CSP), and solar thermal. Although all three aim to maximize the capture of usable energy from sunlight, their approaches are radically different. PV cells convert the energy of sunlight directly into electricity, while CSP converts concentrated heat energy from sunlight into electricity. Solar thermal uses sunlight's energy to heat water and buildings with no intermediate conversion to electricity.

Production up, costs down

Manufacturing continues to expand in the United States and worldwide. As the industry has grown over the past 25 years, the cost of PV has declined by several orders of magnitude. The PV industry estimates that the system price paid by the end user will be $3–4 per watt in 2010 [compared with $7 to $10 per watt in 2004]. This power is being produced in the middle of the day, when electricity costs and needs are the highest. Furthermore, locally installed PV reduces stress on the transmission and distribution system.

> '*The domestic photovoltaic industry will provide up to 15 percent . . . of the new U.S. peak electricity generating capacity expected to be required in 2020.*'

Solar energy is especially well suited for heating water, a task that requires 15–20 percent of a home's total energy consumption. Solar water heaters can provide 50–90 percent of that hot water, and their original cost can be recovered through energy bill savings over the course of 4–7 years. Already, the most economical way to heat a pool is with solar. Today, 20 percent of all units sold to heat pools are solar. The potential value of the technology is shown in Israel, where solar hot water heaters displace 6 percent of the country's total electricity consumption.

Some of the distributed CSP technologies, such as roof-mounted systems, solar dishes, and concentrating PV, will also play a role in America's energy future.

> *A recent* Newsweek *poll recorded that 84 percent of Americans desire more federal investment in solar and wind energy.*

According to the U.S. Photovoltaic Industry Roadmap, "The domestic photovoltaic industry will provide up to 15 percent (about 3,200 MW or 3.2 GW) of the new U.S. peak electricity generating capacity expected to be required in 2020." Cumulative U.S. PV shipments, both domestic and abroad, should stand at about 36 GW at that time. By 2020, if current growth trends are sustained, over 150,000 Americans will be employed in the PV industry.

Environmentally friendly

PV's ability to generate electricity while producing no atmospheric emissions or greenhouse gases marks it as a technology of choice. A 2.5 kW system (enough to power a typical home) covers less than 400 square feet of rooftop. Over the course of a year, it saves about the same amount of carbon dioxide that a car emits during that period.

The public recognizes, now more than ever, that it is time to embrace renewable technologies. A recent *Newsweek* poll recorded that 84 percent of Americans desire more federal investment in solar and wind energy. In November 2001, 73 percent of voters in San Francisco supported a $100 million bond to place solar on buildings in their city. The Sacramento Municipal Utility Disrict (SMUD) has a waiting list of people who wish to install solar on their rooftops and has placed over 10 MW of systems in service. Home builders are partnering with companies in making solar a standard feature. In certain locations in California, Home Depot is selling complete solar energy systems. This demand will spread. Building-integrated PV systems are also gaining momentum.

Zero net energy buildings

The Department of Energy's Office of Energy Efficiency and Renewable Energy, under the capable leadership of David Garman, is embarking on an initiative called zero net energy buildings (ZEB). As its name implies, the program's goal is to produce buildings that on net consume no energy. The zero net energy building will incorporate various technologies, including photovoltaics and solar thermal, to produce the home of the future. To help win the trust of the public, the federal government could play an important role by selecting certain government construction projects to be designated as zero net energy buildings.

The math shows that many small solar generating systems distributed throughout a power grid reduce the need for traditional power facilities. Don Osborn, the superintendent of renewable generation of SMUD, uses the following statistic: If every new home in California placed a 2.2 kW PV system on its rooftop, the equivalent of a 500 MW power plant would be displaced.

The shift to ZEB will benefit far more than the solar industry. Consumers will receive lower electricity bills and enjoy a cleaner environment. A more distributed electricity system will make us less vulnerable to terrorist attacks and benign disruptions. Stress on the transmission grid will be relieved. Energy independence could be achieved.

Even with a more distributed electricity model, central power stations are not likely to disappear. Rather, over time, the central energy sources should become cleaner. Among these, CSP technologies have a bright future. Congress has requested that the administration prepare a report on how to produce 1,000 MW of CSP in the Southwest. . . .

The United States boasts some of the best solar resources on earth and a people dedicated to a cleaner, more sustainable future for their children and grandchildren. If we match this boast with the groundbreaking achievements in solar energy lying within our grasp, the twenty-first century can enter the annals as the solar century.

9

Solar and Wind Power Are Unproductive and Environmentally Harmful

Paul K. Driessen

Paul K. Driessen is a senior fellow with the Committee for a Constructive Tomorrow and Center for the Defense of Free Enterprise, nonprofit public policy institutes that focus on energy, the environment, economic development and international affairs.

Environmentalists try to prevent oil, gas, coal, and nuclear power development in the United States by arguing that alternatives such as solar collectors and wind mills are superior. In reality, these sources of "green" energy can only provide a fraction of the energy needed to keep the American economy strong. And if solar panels and wind farms were built in the numbers necessary to supply more power, they would create visual blight and environmental harm nearly equal to traditional energy sources.

A cacophony of calumny has greeted suggestions that America begin drilling in Alaska's Arctic National Wildlife Refuge [ANWR], Outer Continental Shelf, and other public lands, in search of oil and natural gas, to ease our spreading energy crisis and help rein in prices. The Sierra Club, Senator Barbara Boxer (D-California), and others say drilling is unacceptable and multiple use is "out of the mainstream" of American thought.

Paul K. Driessen, "The False Promise of Renewable Energy," www.heartland.org, May 2001. Copyright © 2001 by the Heartland Institute. Reproduced by permission.

Accompanying the chorus of condemnation for fossil fuels, predictable paeans of praise are being warbled for renewable fuels, as the only true, "appropriate" path to energy security. But all renewables are not equal in the eyes of the environmentalists.

- Hydroelectric power dams up streams, interferes with migratory fish, and impairs the "wilderness experience" of river rafters.
- Burning wood causes serious air quality problems (hydrocarbons and soot, in particular) and requires that trees be cut down, a definite no-no among greens.
- Geothermal suffers from the insurmountable problems that natural heat sources are few and far between . . . and located near magnificent wonders like Yellowstone, Lassen Volcanic, and Hawaii Volcanoes National Parks.
- And soaring natural gas prices have sent fertilizer prices into the stratosphere, making biomass more costly to grow and ship than it will fetch on the open market.

So these renewable fuels are no longer quite politically correct, environmentally defensible, economically possible, or socially "responsible."

That leaves us with but two alternatives to the nuclear and fossil fuels that have powered our progress and prosperity for decades. Solar and wind power have long been touted as the answer to prayers for inexhaustible, non-polluting energy sources. But can they live up to their advance billing?

Vast energy farms

Even today, their total contribution stands at less than 0.5 percent of America's energy needs. Aside from their still-high cost, the primary drawbacks for solar and wind power are that they are intermittent; there is no economic way to store the electrical energy for use at night, on cloudy or windless days, and during peak usage hours; and their environmental impacts are significant and negative.

Producing 50 megawatts of electricity using a gas-fired generating plant requires between 2 and 5 acres of land. Getting the same amount from photovoltaics means covering some 1,000 acres with solar panels (assuming a very optimistic 10 watts per square meter ($W/m2$) or 5 percent peak efficiency), plus access for trucks to clean the panels. Using the sun to meet California's energy needs would require paving over tens of thousands of acres of desert habitat, sacrificing what the Wilderness Society calls "some of the most beautiful landscapes in America," and with it their resident plant and animal life.

A 50 mw wind facility requires even more land: some 4,000 acres (assuming an optimistic 6 W/m2). Even wind power's most ardent supporters grudgingly admit that the notion of thousands of these "futuristic looking" (a euphemism for ugly) towers looming 100 to 200 feet above the rolling hills is not something they yearn to have in their own back yards.

> *There is no economic way to store the electrical energy for use at night, on cloudy or windless days, and during peak usage hours.*

Wind facilities in Texas and California have been called a "visual blight." Residents near Texas' Altamont Pass facility say noise from the turbines is "unbearable."

Vocal California activists have railed for years against off-shore oil platforms on the horizon off Santa Barbara. Are we to believe they will find vast "energy farms" of giant windmills more tolerable?

Killing birds

Noise and visual blight are only the beginning of wind power's adverse environmental consequences, however. Even the relatively small number of wind turbines that exist today kill some 500 hawks, vultures, eagles, and other raptors every year, along with thousands of other birds. The Sierra Club has aptly called them "Cuisinarts of the air," and the U.S. Fish and Wildlife Service has actually suggested that wind farm operators might be prosecuted and jailed for killing federally protected birds. How's that for an incentive to get into the business?

On a national scale, the environmental impacts of solar and wind power become truly staggering. Former Deputy Energy Secretary Ken Davis has calculated that, to produce the 218 gigawatts of additional electricity America will need by 2010, using only wind or solar power, we would have to blanket 9,400,000 acres with windmills or solar panels. That's nearly 10 percent of California . . . an area equal to Connecticut, Delaware, and Massachusetts combined!

Perhaps some photovoltaic panels will be located on roofs, and some of the land in between windmills can be used for farming and grazing (which [environmentalists] also dislike). However, the total acreage affected by these "Earth-friendly" energy sources would still run into the millions. By contrast, developing the 6 to 16 billion barrels of recoverable oil estimated to be in ANWR's distant coastal plain would disturb only 2,000 acres.

Our nation—and California in particular—has lived in a world of energy alchemy and make-believe for long enough. It's high time we recognized there is no free lunch or magic elixir. Tough decisions must be made if progress, prosperity, and opportunity are not to become only a dim memory.

10

The Use of Wind Energy Should Be Increased

American Wind Energy Association

The American Wind Energy Association is a national trade association that represents wind power plant developers, wind turbine manufacturers, and others involved in the wind industry.

Wind turbines are simple, effective devices that only need the blowing wind to produce power. They can be located individually on private property or grouped together on wind farms. This green wind energy alternative can replace dirty, polluting coal and oil power plants resulting in fewer cases of asthma and lung cancer, less acid rain, and relief from global warming.

Wind energy is a form of solar energy, created by circulation patterns in the Earth's atmosphere that are driven by heat from the sun.

People have made use of wind energy for thousands of years, fashioning sails and attaching them to boats for transportation or to wind mills to grind grain. The energy that the wind contains can either be used directly, as in these examples, or it can be converted into that high-value, highly flexible and useful form of energy we call electricity.

Perhaps the simplest way to describe a wind-electric turbine generator (or "wind turbine," as it is usually called) is to say that it works just like a hydroelectric generator. At hydropower stations throughout the U.S. and the world, the energy that is contained in falling or flowing water is used to spin the rotor of a turbine (a rotor that looks quite a bit like an everyday electric fan), and the turbine rotor drives the shaft of a generator to produce electricity.

Wind energy actually works in very similar fashion, especially similar to "run-of-the-river" hydro stations that make use of the flowing water in a river or stream. In the case of wind, of course, the "river" is an invisible one made of air, but the principle is the same. As the air flows past the rotor of a wind turbine (a rotor that looks a lot like an airplane propeller), the rotor spins and drives the shaft of an electric generator.

What is different about wind?

First and most importantly, the fluid (air) that drives the rotor is much less dense than water, and so the diameter of the rotor must be much larger than the rotor of a hydro turbine. A hydro turbine capable of generating one megawatt (MW) of power would be several feet in diameter—a 1-MW wind turbine's rotor would be roughly 175 feet across.

> *About one-third of the U.S. (an area stretching from Minnesota to Texas to Wyoming) has enough wind almost everywhere to generate electricity economically.*

Second, wind energy is available over a much larger geographical range than hydropower—about one-third of the U.S. (an area stretching from Minnesota to Texas to Wyoming) has enough wind almost everywhere to generate electricity economically, and there are many hills and passes in other states that are windy enough as well. Altogether, 46 of the 50 states have some wind resources that could be developed.

Wind turbines come in all sizes, from those with rotors measuring a few feet across (often used for battery charging on sailboats or vacation homes) to those with rotors hundreds of feet in diameter (used to generate "bulk" electricity that is fed into the utility transmission and distribution system). Turbine subsystems include:

- a rotor, or blades, which convert the wind's energy into rotational shaft energy;
- a nacelle [utility box] containing a drive train, usually in-

cluding a gearbox* and a generator;
- a tower, to support the rotor and drive train; and
- electronic equipment such as controls, electrical cables, ground support equipment, and interconnection equipment.

* Some turbines operate without a gearbox.

Household wind systems have rotors up to perhaps 25 feet in diameter, and can be an attractive choice [for those who] live in a windy area or have high electricity prices. . . .

Good wind speeds are important! The energy that the wind contains is a function of the cube of its speed. This means that a site with 12-mph average winds has more than 70% more energy than a site with 10-mph average winds.

Reduce the use of fossil fuels

Utility-scale wind systems typically generate electricity at lower cost—as low as 3–6 cents per kilowatt-hour [as opposed to 10 cents per kh for home systems]. Most regions of the U.S. are served by "power pools" of utilities that join together to generate electricity and transmit it to where it is needed. The name "power pool" is an apt one—electricity coming from many different sources (a coal-fired power plant, a hydro plant, and others) flows into a "pool" from which it is distributed to thousands of end users. A power pool can easily absorb the electricity from a wind plant and add it to all the rest. Wind plants could be installed in many parts of our country, providing income, jobs, and electricity for homes and businesses.

> *Development of 10% of the wind potential in the 10 windiest U.S. states would provide more than enough energy to displace emissions from the nation's coal-fired power plants.*

Experience also shows that wind power can provide at least up to a fifth of a system's electricity, and the figure could probably be higher. Wind power currently provides more than 20% of the electricity distributed by Energia Hidroelectrica de Navarra, the regional electric utility of the industrial state of Navarra in northern Spain. In Denmark, wind supplies 20% of the nation's electricity. If wind energy in the U.S. were combined with serious efforts to increase energy efficiency, we could substantially reduce our national use of fossil fuels to generate electricity.

Today, utility-scale wind turbines worldwide total over 30,000 megawatts of generating capacity. Yet this is but a tiny fraction of wind's potential. A recent study performed by Denmark's BTM Consult for the European Wind Energy Association and Greenpeace found that by the year 2017, wind could provide 10% of world electricity supplies, meeting the needs of 500 million average European households.

One key issue for utility-scale wind plants that must be resolved in the coming years is transmission line capacity. Utility transmission lines are like a "pipeline" that is needed to carry wind-generated electricity from the vast and sparsely populated areas of the Great Plains, where the wind is most abundant, to large cities like Minneapolis, Milwaukee, Chicago, and Dallas where demand for electricity is high. At the moment, there are not many transmission lines that connect cities with the windiest parts of the plains.

Pollution-free power

Wind energy is a particularly appealing way to generate electricity because it is essentially pollution-free. More than half of all the electricity that is used in the U.S. is generated from burning coal, and in the process, large amounts of toxic metals, air pollutants, and greenhouse gases are emitted into the atmosphere.

Development of 10% of the wind potential in the 10 windiest U.S. states would provide more than enough energy to displace emissions from the nation's coal-fired power plants and eliminate the nation's major source of acid rain; reduce total U.S. emissions of carbon dioxide (the most important greenhouse gas) by almost a third and world emissions of CO_2 by 4%; and help contain the spread of asthma and other respiratory diseases aggravated or caused by air pollution in this country. If wind energy were to provide 20% of the nation's electricity—a very realistic and achievable goal with the current technology—it could displace more than a third of the emissions from coal-fired power plants, or all of the radioactive waste and water pollution from nuclear power plants.

Also, wind farms can revitalize the economy of rural communities, providing steady income through lease or royalty payments to farmers and other landowners. Although leasing arrangements can vary widely, a reasonable estimate for income to a landowner from a single utility-scale turbine is about $2,000 a year or more, depending on the wind resource, the size of the turbine, and other factors. For a 250-acre farm, with income from wind at about $55 an acre, the annual income from a wind lease would be $14,000, with no more than 2–3 acres removed from production. Farmers can grow crops or raise cattle next to the towers. Wind farms may extend over a large geographical area, but their actual "footprint" covers only a very small portion of the land, making wind development an ideal way for farmers to earn additional income. In west Texas, for example, farmers are welcoming wind, as lease payments from this new clean energy source replace declining payments from oil wells that have been depleted.

Farmers are not the only ones in rural communities to find that wind power can bring in income. In Spirit Lake, Iowa, the local school is earning savings and income from the electricity generated by a turbine. In the district of Forest City, Iowa, a turbine recently erected as a school project is expected to save $1.6 million in electricity costs over its lifetime.

Greater use of wind energy means a cleaner environment with healthier air, and more income to landowners and economically depressed counties and communities in the Great Plains. It means relying more on an energy source whose "fuel" is free and will never be exhausted or embargoed.

11

The Use of Wind Energy Should Not Be Increased

Eric Rosenbloom

Eric Rosenbloom has twenty years experience in the publishing industry. He has written Internet articles on alternative energy, vegetarianism, and hemp fibers.

People in the United States and Europe have spent hundreds of millions of dollars in order to harness the energy of the wind. From South Dakota to Denmark, experiments with wind farms have shown, however, that wind power does not work as promised. The turbines used in windmills are extremely loud when turning at speeds of over one hundred miles per hour and create intolerable noise pollution. Windmills create visual blight, kill birds, and do not produce nearly enough power to justify their expense. While alternatives to oil, coal, and nuclear power are needed, windmills are not the answer to the worlds' ever-increasing appetite for electrical energy.

Wind power promises a clean and free source of electricity. It will reduce our dependence on imported fossil fuels and reduce the output of greenhouse gases and other pollution. Many governments are therefore promoting the construction of wind "farms," encouraging private companies with generous subsidies and regulatory support and by requiring utilities to buy from them. The U.S. Department of Energy (DOE) aims to see 5% of our electricity produced by wind turbine in 2010. Energy companies are eagerly investing in wind power, finding the arrangement quite profitable.

A little research reveals that wind power does not in fact

live up to the claims made by its advocates, that its impact on the environment and people's lives is far from benign, and that with such a dismal record the money spent on it could be much more effectively directed. . . .

In 1998, Norway commissioned a study of wind power in Denmark and concluded that it has "serious environmental effects, insufficient production, and high production costs."

Denmark (population 5.3 million) has over 6,000 turbines that produced electricity equal to 19% of what the country used in 2002. Yet no conventional power plant has been shut down. Because of the intermittency and variability of the wind, conventional powerplants must be kept running at full capacity to meet the actual demand for electricity. They cannot simply be turned on and off as the wind dies and rises, and such inefficient operation would actually increase their output of pollution and carbon dioxide (the primary "greenhouse" gas). So when the wind is blowing just right for the turbines, the power they generate is usually a surplus and sold to other countries, or the turbines are simply shut off. As noted by *The Wall Street Journal Europe*, the Copenhagen newspaper *Politiken* reported that in 1999 wind actually met only 1.7% of Denmark's total demand.

(Extrapolating the latter figure to the U.S. (population 290,000,000, per-capita electricity use twice Denmark's), almost 2 million wind towers would be needed to reach the DOE's 5% goal.)

> **❝** *In high winds, ironically, the turbines must be stopped because they are easily damaged.* **❞**

Denmark is just dependent enough on wind power that when the wind is not blowing right they must import electricity. In 2000 they imported more electricity than they exported. And added to the Danish electric bill is the subsidy that supports the private companies building the wind towers. Danish electricity costs for the consumer are the highest in Europe. . . .

The head of Xcel Energy in the U.S., Wayne Brunetti, has said, "We're a big supporter of wind, but at the time when customers have the greatest needs, it's typically not available." A study by the U.K. Department of Trade and Industry (DTI) in

1998 found that wind turbines produced on average less than 25% of their theoretical . . . capacity over a year. Throughout Europe, the average was less than 20%. . . . In California, the average is 20%. . . . The claimed generating capacity only occurs during 100% ideal conditions, typically a sustained wind speed over 30 mph. As the wind slows, electricity output falls off sharply.

In high winds, ironically, the turbines must be stopped because they are easily damaged. Build-up of dead bugs has been shown to halve the maximum power generated by a wind turbine, reducing the average power generated by 25% and more. Build-up of salt on offshore turbine blades similarly has been shown to reduce the power generated by 20%–30%.

Canceling plans for wind farms

According to *Windpower Monthly*, an industry journal, both Denmark and Germany are taking action against the high cost of insurance and premature failure of large turbines. In Princeton, Massachusetts, three of a developer's eight towers have long stood idle because of lightning damage.

The Danish government has cancelled plans for three offshore windfarms planned for 2008 and has scheduled the withdrawal of subsidies from existing sites. Development of onshore wind plants in Denmark has effectively stopped. Spain began withdrawing subsidies in 2002. Germany is considering ending subsidies to wind power. Switzerland also is cutting subsidies as too expensive for the lack of significant benefit. It is reported that California will no longer seek new installations and will instead upgrade those that already exist (which also face the problem of high rates of bird deaths). Ireland in December 2003 halted all new wind-power connections to the national grid, because of the serious instability they cause.

[Because of government regulations in] Germany, utilities are forced to buy renewable energy at sometimes more than 10 times the cost of conventional power, in France 3 times. In the U.K. [United Kingdom], the *Telegraph* has reported that rather than providing cheaper energy, wind power costs the electric companies £50 per megawatt-hour, compared to £15 for conventional power. . . . The wind industry is worried that the U.K., too, is starting to see that it is only subsidies and requiring utilities to buy a certain amount of "green" power that prop up the wind towers and that it is a colossal waste of resources. The British Wind Energy Association [BWEA] has even resorted to threatening prominent opponents as more projects are successfully blocked. Interestingly, long-term plans for energy use and emissions reduction by both the U.K. and the U.S. governments do not mention wind.

In the U.S., wind power plants enjoy tax credits and accelerated depreciation. These costs could be justified if large-scale wind power provided real benefits, but, as this essay shows, it does not.

Cannot keep up with demand

Installation of wind towers cannot hope to keep up with the continuing increase of energy use. . . . In Germany (population 82.5 million), a group of engineers and other scientists pointed out in 1998 that more than 5,000 turbines provided less than 1% of their electricity. At the end of 2002, the figure was about 14,000 turbines, one-third of the world's total wind power, producing just over 2% of Germany's electricity. A study in *Science* (November 1, 2002) figured that world energy demand will increase severalfold by midcentury; the researchers found that there is no viable technology today for significantly replacing even today's fossil fuel use.

In the U.K. (population 60 million), 1,010 wind turbines produced 0.1% of their electricity in 2002, according to the Department of Trade and Industry. The government hopes to increase the use of renewables to 10.4% by 2010 and 20.4% by 2020, requiring many tens of thousands more towers. As demand will have grown, however, even more turbines would be required. In California (population 35 million), according to the state energy commission, 14,000 turbines produced half of one percent of their electricity in 2000. Extrapolating this record to the U.S. as a whole, and without accounting for an increase in energy demand, 1.2 million wind towers (costing about $2.4 trillion) would be necessary to meet the DOE's goal of a mere 5% of the country's electricity from wind by 2010.

The DOE says there are 18,000 square miles of good wind sites in the U.S., which with current technology could produce 20% of the country's electricity. (This rosy plan, based on only the wind industry's sales brochures, as well as on a claim of electricity use that is only three-quarters of the actual use in 2002, would require "only" 142,060 1.5-MW [megawatt] towers.) They also explain, "If the wind resource is well matched to peak loads, wind energy can effectively contribute to system capacity." That's a big *if*—counting on the wind to blow exactly when demand rises, especially if you expect the wind to cover 20% (or even 5%) of that demand. As in Denmark, you would quickly learn that the prudent thing to do is to look elsewhere first in meeting the load demand. And we'd be stuck with a lot of generally unhelpful hardware covering every windy spot in the U.S., while the developers would be looking to put up yet more to make up for and deny their failings.

As in Denmark, the electricity from those towers—no matter how many—would be too variable to provide the constant supply that the grid demands. (More than two-thirds of the time, a tower is effectively not generating electricity at all.) They have no effect on established electricity generation, energy use, or continuing pollution. Christopher Dutton, the CEO of Green Mountain Power, a partner in the Searsburg wind farm in Vermont and an advocate of alternative energy sources, has said (in an interview with Montpelier's *The Bridge*) that there is no way that wind power can replace more traditional sources, that its value is only as a supplemental source that has no impact on the base load supply. "By its very nature, it's unreliable," says Jay Morrison, senior regulatory counsel for the National Rural Electric Cooperative Association.

As Country Guardian, a U.K. conservation group, puts it, wind farms constitute an *increase* in energy supply, not a replacement. They do not reduce the costs—environmental and economic—of other means of energy production. If wind towers do not reduce conventional power use, then their manufacture, transport, and construction only increases the use of dirty energy. The presence of "free and green" wind power may even give people license to use *more* energy.

Construction destruction

Pictures from the energy companies show slim towers rising cleanly from the landscape or hovering faintly in the distant haze, their presence modulated by soft clouds behind them. But a 200- to 300-foot tower supporting a turbine housing the size of a bus and three 100- to 150-foot rotor blades that spin at 100 mph or more requires, for a start, a large and solid foundation. On a GE 1.5-MW tower, the turbine housing (nacelle) weighs over 56 tons, the blade assembly weighs over 36 tons, and the whole tower totals over 163 tons.

As FPL (Florida Power & Light) Energy says, "a typical turbine site takes about a 42×42-foot-square graveled area." Each tower (and a site needs at least 15–20 towers to make investment worthwhile) requires a huge hole filled with tons of steel-reinforced concrete. According to Country Guardian, the hole is large enough to fit three double-decker buses. It may be 30 feet deep or more and contain more than 100,000 cubic feet of concrete (production of which is a major source of CO_2). On mountain ridges and many other locations, it would be necessary to blast into the bedrock, possibly disrupting the water sources for wells downhill. Construction at a site on the Slieve Aughty range in Ireland in October 2003 caused a 2.5-mile-long bog slide. (Building on peat bogs is recognized as a serious disruption of an important carbon sink; the Royal Society for the Protection of Birds opposes wind development on the Scottish island of Lewis because the turbines would take 25 years to save the amount of carbon that their construction will release from the peat.)

> *Especially vulnerable are large birds of prey that like to fly in the same sorts of places that developers like to construct wind towers.*

FPL Energy also says, "although construction is temporary [a few months], it will require heavy equipment, including bulldozers, graders, trenching machines, concrete trucks, flatbed trucks and large cranes." Getting all the equipment, as well as the huge tower sections and rotor blades, into an undeveloped area requires the construction of wide straight roads. Many an ancient hedgerow in England has been sacrificed for access to project sites.

The destructive impact that such construction would have, for example, on a wild mountain top, is obvious. Erosion, disruption of water flow, and destruction of wild habitat and plant life would continue with the presence of access roads, power lines, transformers, and the tower sites themselves. For better wind efficiency, each tower requires trees to be cleared. Vegetation would be kept down with herbicides, further poisoning the soil and water. Each tower should be at least 5–7 times the rotor diameter from neighboring towers and trees for optimal performance. For a tower with 35-meter rotors, that is 1200–

1600 feet, a quarter to a third of a mile. A site on a forested ridge would require clearing 45–90 acres per tower to operate optimally (although only 4–5 acres per tower, the towers spaced every 500 feet or so, is typical).

GE boasts that the span of their rotor blades is larger than the wingspan of a Boeing 747 jumbo jet. The typical 1.5-MW tower is two stories higher than the Statue of Liberty, including its base and pedestal. The editor of *Windpower Monthly* wrote in September 1998, "Too often the public has felt duped into envisioning fairy tale 'parks' in the countryside. The reality has been an abrupt awakening. Wind power stations are no parks." They are industrial and commercial installations. They do not belong in wilderness areas. As the U.K. Countryside Agency has said, it makes no sense to tackle one environmental problem by instead creating another.

In Vermont, billboards are banned from the highways, and development—especially at sites above 2500 feet—is subject to strong environmental laws, yet the state absurdly supports the installation of wind farms as a desirable trade-off, ignoring wind's dismal record. . . .

Even if one thinks that jumbo-jet-sized wind towers dominating every ridge line in sight like a giant barbed-wire fence is a beautiful thing, many people are drawn to wild places to avoid such reminders of human industrial might. Many communities depend on such tourists, who will now seek some other—as yet unspoiled—retreat.

Killing birds and bats

Yes, the towers kill (and maim) birds and bats. The Danish Wind Industry Association, for example, admits as much by pointing out that so do power lines and automobiles. (Their argument follows the aesthetic one that the landscape is already blighted in many ways, so why not blight it some more?) The industry claims that moving from lattice-work towers, which provided roosting and nesting platforms, to solid towers, as well as larger lower-rpm blades, solved the problem, and that studies find very few dead birds around wind turbines. They ignore the facts that the larger blades are in fact slicing the air faster—over 100 mph at their tips—and that scavengers remove most injured and dead birds before researchers arrive for their periodic surveys.

Especially vulnerable are large birds of prey that like to fly in the same sorts of places that developers like to construct wind towers. Fog—a common situation on mountain ridges—aggravates the problem for all birds. Guidelines from the U.S. Fish and Wildlife Service [FWS] state that wind towers should not be near wetlands or other known bird or bat concentration areas or in areas with a high incidence of fog or low-cloud ceilings, especially during spring and fall migrations. It is illegal in the U.S. to kill migratory birds. The FWS has prevented any expansion of the several Altamont Pass wind plants in California.

A 2002 study in Spain estimated that 11,200 birds of prey (many of them already endangered), 350,000 bats, and 3,000,000 small birds are killed each year by wind turbines and their power lines. An analysis of government reports found that it is officially recognized (and obscured) that on average a single turbine tower kills 20–40 birds each year. The U.S. Fish and Wildlife Service estimates that European wind power kills 37 birds per turbine each year. *Windpower Monthly* reported in October 2003 that the shocking number of bats being killed by wind towers in the U.K. is causing trouble for developers.

Jobs, taxes, and property values

Despite the energy industry's claim that wind farms create jobs ("revitalize struggling rural communities," says enXco), the fact is that, after the few months of construction, a typical large wind farm requires just one maintenance worker.

The energy companies also claim that they increase the local tax base. But that is more than offset by the loss of open land, the loss of tourism, the stagnation or decrease in property values throughout a much wider area, the tax credits such developments typically enjoy, and the taxes and fees consumers must pay to subsidize the industry. A few people get more money from leasing their land for the towers (until the developer starts withholding it for some small-print reason, or even disappears after the tax advantages slow down), but that's the opposite of an argument for the general good. Even surveys by wind supporters show that a quarter to a third of visitors would no longer come if wind turbines were installed. That is a huge loss.

Wind advocates insist that property values are not affected by nearby industrial turbines, because there will always be a buyer as it's just a question of taste. That is small comfort to those who already own homes near potential wind-plant sites but whose taste militates against rattling windows and humming walls, flickering lights, 100-foot blades spinning overhead, and giant metal towers and supply roads where once were trees and moose trails. The next section provides some examples of the serious effect of industrial wind plants on property values.

Noise

Noise is another big problem the industry claims to have solved. Indeed, new turbines have quieter gears (though still far from silent), but the problem of 100-foot rotor blades chopping through the air at over 100 mph is insurmountable. (A 35-meter [115-foot] blade turning at 15 rpm is travelling 123 mph at the tip, at 20 rpm 164 mph.) Every time each rotor passes the tower, the compression of air produces a deep resonating thump. The resulting sound of several towers together has been described to be as loud as a motorcycle, like aircraft continually passing overhead, a "brick wrapped in a towel turning in a tumbledrier," "as if someone was mixing cement in the sky." There is a penetrating low-frequency aspect to the noise, a thudding vibration, much like the throbbing bass of a neighboring disco, that travels much farther than the usually measured "audible" noise.

The beat is often close to the human heart rate, and people have complained that it causes anxiety and nausea. The only way to reduce it is to reduce the efficiency of the electricity production, i.e., reduce the illusion of profitability. It can't be done.

The European Union published the results of a 5-year investigation into wind power, finding noise complaints to be valid and that noise levels could not be predicted before developing a site. One manufacturer specifies that their turbines not be placed within 2 kilometers (1.25 miles) of any dwelling. The American Wind Energy Association (AWEA) acknowledges that a turbine is quite audible 800 feet away. The National (U.S.) Wind Coordinating Committee states, "wind turbines are highly visible structures that often are located in conspicuous settings . . . they also generate noise that can be disturbing to nearby residents."

Communities in Germany, Wales, and Ireland claim that even 3,000 feet away, the noise is significant. . . . A German study in 2003 found significant noise levels 1 mile away from a 2-year-old wind farm of 17 1.8-MW turbines, especially at night. In mountainous areas the sound echos over larger distances. In Vermont, the director of Energy Efficiency for the Department of Public Service has said that the noise from the

11 550-KW Searsburg turbines is significant a mile away. Residents 1.5 and even 3 miles downwind in otherwise quiet rural areas suffer significant noise pollution. A criminal suit has been allowed to go forward in Ireland against the owner and operator of a wind plant for noise violations of their environmental law. Also in Ireland, a developer has been forced to compensate a homeowner for loss of property value, and many people have had their tax valuation reduced. In the Lake District of northwest England, a group has sued the owner and operator of the Askam wind plant, claiming it is ruining their lives.

> *[The] sound of several towers together has been described to be as loud as a motorcycle, like aircraft continually passing overhead, a 'brick wrapped in a towel turning in a tumbledrier.'*

In January 2004, a couple was awarded 20% of the value of their home from the previous owners who did not tell them the Askam wind plant was about to be constructed 1800 feet away: "because of damage to visual amenity, noise, pollution, and the irritating flickering caused by the sun going down behind the moving blades." The towers of this plant are only 40 meters (130 feet) high, with the rotors extending the height a further 24 meters (75 feet). Steve Molloy of West Coast Energy responded that loss of value of a property, although unfortunate, was not a material planning consideration and did not undermine the industry's argument that the benefits of sustainable energy outweighed the objections.

Don Peterson, senior director of Madison Gas & Electric, which operates 31 wind towers in Kewaunee County, Wisconsin, similarly dismisses complaints, saying that most people, but not all, will get used to the sound of the machines. "Like any noise, if you don't like it, your brain is going to focus on it," he has comfortingly said. It has been reported that one of the farmers who leases land for the wind towers had to buy the neighbors' property because of the problems (not just noise but also flicker and lights at night). On January 6, 2004, the *Western Morning News* of Devon published three articles about noise problems, particularly the health effects of low-frequency noise, from wind turbines. . . .

Other problems

The industry recognizes that the flicker of reflected light on one side and shadow on the other drives people and animals crazy. And at night, the towers must be lighted, which the AWEA describes as a serious nuisance, destroying the dark skies that many people in rural areas cherish. Red lights are thought to attract night-migrating birds.

Ice is another problem. It builds up when the blades are still and gets flung off—as far as 1,500 feet—when they start spinning.

The International Association of Engineering Insurers warns of fire: "Damage by fire in wind turbines is usually caused by overheated bearings, a strike of lightning, or sparks thrown out when the turbine is slowing down. . . . Even the smallest spark can easily develop into a large fire before discovery is made or fire-fighting can begin."

Lightning destroys many towers by causing the blade coatings to peel off, rendering them useless. The towers are subject to metal fatigue, and the resin blades are easily damaged, even by wind. In Wales, Spain, and Germany, parts and whole blades have torn off, flying as far as 1,200 feet and through the windows of homes. Whole towers have collapsed in Germany (as recently as 2002) and the U.S.

Money better spent elsewhere

These negative aspects should of course be weighed against industrial wind power's benefits. However, there are none.

It is wise to diversify the sources of our energy. The money and legislative effort invested in large-scale wind generation could be spent much more effectively to achieve the goal of reducing our use of fossil and nuclear fuels.

As an example, Country Guardian calculates that for the U.K. government subsidy towards the construction of one wind turbine, they could insulate the roofs of almost 500 houses that need it and save in two years the amount of energy the wind turbine might produce over its lifetime.

Country Guardian also calculates that if every light bulb in the U.K. were switched to a more efficient one, the country could shut down an entire power plant—something even Denmark, with wind producing up to 20% of their electricity, is not able to do. According to solar energy consultant and retailer Real Goods, if every household in the U.S. replaced one incan-

descent bulb with a compact fluorescent bulb, one nuclear
power plant could be closed. John Etherington claims that
switching the most-used bulb in every house of the U.K. would
save as much as the entire output of all existing and proposed
on-shore wind plants in that country.

The U.N.-sponsored Intergovernmental Panel on Climate
Change has stated that simple voluntary energy-efficiency improvements in buildings will reduce world energy use 10%–
15% by 2020. The BWEA itself says that the cost of saving energy is less than half the cost of producing it. According to the
California Power Authority, ignoring the subsidies that lower
the market price of wind-generated electricity, conservation
costs exactly the same per KW-h as wind power. . . .

Wind farms do not bring about any reduction in the use of
conventional power plants. Subsidizing the upgrading of
power plants to be more efficient and cleaner would actually
do something rather than simply support the image of "green"
power that energy companies profit from while in fact doing
nothing to reduce pollution or fuel imports. An April 2000 E.U.
[European Union] report found that, using existing technology, increased efficiency could decrease energy consumption
by more than 18% by 2020. In the U.S., 7.34% of the electricity generated is lost in transmission; investing in improved
power lines would save more energy than could ever be generated by wind turbines.

Electricity represents only 39% of energy use in the U.S. (in
Vermont, 20%; and only 1% of Vermont's greenhouse gas
emissions is from electricity generation). Pollution from fossil
fuels also comes from transportation (cars, trucks, and aircraft)
and heating. Demanding better gas mileage in cars, including
pickup trucks and SUVs, promoting rail for both freight and
travel, and supporting the use of biodiesel (for example, from
hemp) would make a huge impact on pollution and dependence on foreign oil, whereas wind power makes none. Hybrid
gas-electric cars already use 60% less gasoline than average conventional new cars in the U.S.

Wind-power advocates often propose that wind turbines
can be used to manufacture hydrogen for fuel cells. This is an
admirable plan but so far in the future that it only serves to un-
derscore the fact that there is no good reason for current construction of large-scale wind power plants.

Alternative energy generation has a place, but it is clearly
impractical to expect wind (or sun) to replace conventional

sources. On a small scale, especially where batteries or flywheels can store the fluctuating production, they can contribute to a single home, school, factory, office building, neighborhood, or even small town's electricity. But on a large scale, so many turbines are necessary to make even a small difference that their negative impact far outstrips even the supposed benefits claimed by their promoters.

We are reminded that there are trade-offs necessary to living in a technologically advanced industrial society, that fossil fuels will run out, that global warming must be slowed, and that the procurement of fossil fuels is environmentally, politically, and socially destructive. Sooner or later the realities of this modern life will have to reach into our own back yards, the commons must be developed for our economic survival, and it would be elitist in the extreme to believe we deserve better. So wilderness areas are sacrificed, rural communities are bribed into becoming live-in (but ineffective) power plants, our government boasts that they are looking beyond fossil fuels (while doing nothing to actually reduce their use), there is no change in our electric bills (they are more likely to go up to support "investment in a greener future"). And at the other end of this trade-off, multinational energy companies reap greater profits.

There are many other alternative sources of energy in development. But wind turbines exist, so they are presented by their manufacturers and managers as *the* solution. True energy independence would be more diverse and noncentralized. But who would profit? How would the lawmakers get their cut?

12

Hydrogen Power Should Be Developed

Jeremy Rifkin

Jeremy Rifkin, president of the Foundation on Economic Trends, is the author of sixteen books on the impact of scientific and technological changes on the economy, the workforce, society, and the environment. His latest book is The Hydrogen Economy: The Creation of the World Wide Energy Web and the Redistribution of Power on Earth.

The control and distribution of fuel and electricity in the modern world is centralized in the hands of a few large corporations. With the advent of hydrogen fuel cells, which generate clean energy from renewable resources, this equation could radically change. With small fuel cell "power plants" available to the world's rich and poor alike in cities and rural villages, energy creation would be in the hands of millions of individuals. This energy "web" would allow economic prosperity to flow along decentralized and democratic lines and would create sustainable power that benefits both society and the environment.

While the fossil-fuel era enters its sunset years, a new energy regime is being born that has the potential to remake civilization along radically new lines—hydrogen. Hydrogen is the most basic and ubiquitous element in the universe. It never runs out and produces no harmful CO_2 emissions when burned; the only byproducts are heat and pure water. That is why it's been called "the forever fuel."

Jeremy Rifkin, "Hydrogen: Empowering the People: A New Source of Energy, If Its Development Follows the Model of the World Wide Web, Offers a Way to Wrench Power from Ever Fewer Institutional Hands," *The Nation*, vol. 275, December 23, 2002, p. 20. Copyright © 2002 by The Nation Magazine/The Nation Company, Inc. Reproduced by permission.

Hydrogen has the potential to end the world's reliance on oil. Switching to hydrogen and creating a decentralized power grid would also be the best assurance against terrorist attacks aimed at disrupting the national power grid and energy infrastructure. Moreover, hydrogen power will dramatically reduce carbon dioxide emissions and mitigate the effects of global warming. In the long run, the hydrogen-powered economy will fundamentally change the very nature of our market, political and social institutions, just as coal and steam power did at the beginning of the Industrial Revolution.

A continuous supply of power

Hydrogen must be extracted from natural sources. Today, nearly half the hydrogen produced in the world is derived from natural gas via a steam-reforming process. The natural gas reacts with steam in a catalytic converter. The process strips away the hydrogen atoms, leaving carbon dioxide as the byproduct.

There is, however, another way to produce hydrogen without using fossil fuels in the process. Renewable sources of energy—wind, photovoltaic, hydro, geothermal and biomass—can be harnessed to produce electricity. The electricity, in turn, can be used, in a process called electrolysis, to split water into hydrogen and oxygen. The hydrogen can then be stored and used, when needed, in a fuel cell to generate electricity for power, heat and light.

Why generate electricity twice, first to produce electricity for the process of electrolysis and then to produce power, heat and light by way of a fuel cell? The reason is that electricity doesn't store. So, if the sun isn't shining or the wind isn't blowing or the water isn't flowing, electricity can't be generated and economic activity grinds to a halt. Hydrogen provides a way to store renewable sources of energy and insure an ongoing and continuous supply of power.

Hydrogen-powered fuel cells are just now being introduced [in 2002] into the market for home, office and industrial use. The major auto makers have spent more than $2 billion developing hydrogen-powered cars, buses and trucks, and the first mass-produced vehicles are expected to be on the road in just a few years.

In a hydrogen economy the centralized, top-down flow of energy, controlled by global oil companies and utilities, would become obsolete. Instead, millions of end users would connect their fuel cells into local, regional and national hydrogen energy webs (HEWs), using the same design principles and smart technologies that made the World Wide Web possible. Automobiles with hydrogen cells would be power stations on wheels, each with a generating capacity of 20 kilowatts. Since the average car is parked most of the time, it can be plugged in, during nonuse hours, to the home, office or the main interactive electricity network. Thus, car owners could sell electricity back to the grid. If just 25 percent of all US cars supplied energy to the grid, all the power plants in the country could be eliminated.

Once the HEW is set up, millions of local operators, generating electricity from fuel cells onsite, could produce more power more cheaply than can today's giant power plants. When the end users also become the producers of their energy, the only role remaining for existing electrical utilities is to become "virtual power plants" that manufacture and market fuel cells, bundle energy services and coordinate the flow of energy over the existing power grids.

To realize the promise of decentralized generation of energy, however, the energy grid will have to be redesigned. The problem with the existing power grid is that it was designed to insure a one-way flow of energy from a central source to all the end users. Before the HEW can be fully actualized, changes in the existing power grid will have to be made to facilitate both easy access to the web and a smooth flow of energy services over the web. Connecting thousands, and then millions, of fuel cells to main grids will require sophisticated dispatch and control mechanisms to route energy traffic during peak and nonpeak periods. A new technology developed by the Electric Power Research Institute called FACTS (flexible alternative current transmission system) gives transmission companies the capacity to "deliver measured quantities of power to specified areas of the grid."

Whether hydrogen becomes the people's energy depends, to a large extent, on how it is harnessed in the early stages of development. The global energy and utility companies will make every effort to control access to this new, decentralized energy network just as software, telecommunications and content companies like Microsoft and AOL Time Warner have attempted to control access to the World Wide Web. It is critical that public institutions and nonprofit organizations—local governments, cooperatives, community development corporations, credit unions and the like—become involved early on in establishing distributed-generation associations (DGAs) in every country. Again, the analogy to the World Wide Web is apt. In the new hydrogen energy era, millions of end users will generate their own "content" in the form of hydrogen and electricity. By organizing collectively to control the energy they produce—just as workers in the twentieth century organized into unions to control their labor power—end users can better dictate the terms with commercial suppliers of fuel cells for lease, purchase or other use arrangements and with virtual utility companies, which will manage the decentralized "smart" energy grids. Creating the appropriate partnership between commercial and noncommercial interests will be critical to establishing the legitimacy, effectiveness and long-term viability of the new energy regime.

> *If just 25 percent of all US cars supplied energy to the grid, all the power plants in the country could be eliminated.*

I have been describing, thus far, the implementation of hydrogen power mainly in industrialized countries, but it could have an even greater impact on emerging nations. The per capita use of energy throughout the developing world is a mere one-fifteenth of the consumption enjoyed in the United States. The global average per capita energy use for all countries is only one-fifth the level of this country. Lack of access to energy, especially electricity, is a key factor in perpetuating poverty around the world. Conversely, access to energy means more economic opportunity. In South Africa, for example, for every 100 households electrified, ten to twenty new businesses are created. Making the shift to a hydrogen energy regime—using renewable resources and technologies to produce the hydrogen—and creating distributed generation energy webs that can connect communities all over the world could lift billions of people out of poverty. As the price of fuel cells and accompanying appliances continues to plummet with innovations and economies of scale, they will become far more broadly available, as was the case with transistor radios, computers and cellular phones. The goal ought to be to provide stationary fuel cells for every neighborhood and village in the developing world.

Renewable energy technologies—wind, photovoltaic, hydro, biomass, etc.—can be installed in villages, enabling them to produce their own electricity and then use it to separate hydrogen from water and store it for subsequent use in fuel cells. In rural areas, where commercial power lines have not yet been extended because they are too expensive, stand-alone fuel cells can provide energy quickly and cheaply.

After enough fuel cells have been leased or purchased, and installed, mini energy grids can connect urban neighborhoods as well as rural villages into expanding energy networks. The HEW can be built organically and spread as the distributed generation becomes more widely used. The larger hydrogen fuel cells have the additional advantage of producing pure drinking water as a byproduct, an important consideration in village communities around the world where access to clean water is often a critical concern.

Were all individuals and communities in the world to become the producers of their own energy, the result would be a dramatic shift in the configuration of power: no longer from the top down but from the bottom up. Local peoples would be less subject to the will of far-off centers of power. Communities would be able to produce many of their own goods and services and consume the fruits of their own labor locally. But, because they would also be connected via the worldwide communications and energy webs, they would be able to share their unique commercial skills, products and services with other communities around the planet. This kind of economic self-sufficiency becomes the starting point for global commercial interdependence, and is a far different economic reality from that of colonial regimes of the past, in which local peoples were made subservient to and dependent on powerful forces from the outside. By redistributing power broadly to everyone, it is possible to establish the conditions for a truly equitable sharing of the earth's bounty. This is the essence of reglobalization from the bottom up.

Empowering the human race

Two great forces have dominated human affairs over the course of the past two centuries. The American Revolution unleashed a new human aspiration to universalize the radical notion of political democracy. That force continues to gain momentum and will likely spread to the Middle East, China and every corner of the earth before the current century is half over.

A second force was unleashed on the eve of the American Revolution when James Watt patented his steam engine, inaugurating the beginning of the fossil-fuel era and an industrial way of life that fundamentally changed the way we work.

The problem is that these two powerful forces have been at odds with each other from the very beginning, making for a deep contradiction in the way we live our lives. While in the political arena we covet greater participation and equal representation, our economic life has been characterized by ever greater concentration of power in ever fewer institutional hands. In large part that is because of the very nature of the fossil-fuel energy regime that we rely on to maintain an industrialized society. Unevenly distributed, difficult to extract, costly to transport, complicated to refine and multifaceted in the forms in which they are used, fossil fuels, from the very beginning, required a highly centralized command-and-control structure to finance exploration and production, and coordinate the flow of energy to end users. The highly centralized fossil-fuel infrastructure inevitably gave rise to commercial enterprises organized along similar lines. Recall that small cottage industries gave way to large-scale factory production in the late nineteenth and early twentieth centuries to take advantage of the capital-intensive costs and economies of scale that went hand in hand with steam power, and later oil and electrification. In the discussion of the emergence of industrial capitalism, little attention has been paid to the fact that the energy regime that emerged determined, to a great extent, the nature of the commercial forms that took shape.

Now, on the cusp of the hydrogen era, we have at least the "possibility" of making energy available in every community of the world—hydrogen exists everywhere on earth—empowering the whole of the human race. By creating an energy regime that is decentralized and potentially universally accessible to everyone, we establish the technological framework for creating a more participatory and sustainable economic life—one that is compatible with the principle of democratic participation in our political life. Making the commercial and political arenas seamless, however, will require a human struggle of truly epic proportions in the coming decades. What is in doubt is not the technological know-how to make it happen but, rather, the collective human will, determination and resolve to transform the great hope of hydrogen into a democratic reality.

13

Hydrogen-Powered Cars May Not Be Feasible

Charles J. Murray

Charles J. Murray is a journalist who has written many technical articles for the Electronic Engineering Times *and is the author of the book* The Supermen: The Story of Seymour Cray and the Technical Wizards Behind the Supercomputer.

While environmentalists, automakers, and the president enthusiastically promote hydrogen fuel cells as the wave of the future, the technology does not exist to produce fuel cells that can power cars. Unless there is an amazing, unforeseen scientific breakthrough, fuel cells will remain prohibitively expensive and impractical for transportation needs. The government, business, and environmental communities are creating public hopes for nonpolluting cars that will not be built for decades, if ever.

When President George W. Bush called on the nation in January [2003] to rally behind the concept of hydrogen-powered cars, automakers and national labs began ratcheting up their efforts to put fuel-cell-powered vehicles on the road by 2010.

But while makers of automobiles—and, by extension, laptop computers and cell phones—may lick their chops at the thought of hydrogen-powered products, experts say the path to fuel cell nirvana could be a long and arduous one, requiring decades to achieve, in some cases.

University professors and researchers with no stake in the success or failure of the technology expect hydrogen-burning

Charles J. Murray, "Fuel Cell R&D Is Far from Easy Street," *Electronic Engineering Times*, May 26, 2003, p. 12. Copyright © 2004 by CMP Media LLC, 600 Community Dr., Manhasset, NY 11030, USA. Reproduced by permission.

fuel cells to make minor inroads in consumer electronics in the next five years, but nary a dent in cars. For fuel cells to reach production automobiles by 2020, these sources said, scientists need a Nobel Prize–winning breakthrough to cut the cost of the technology by a hundredfold or more.

"As far as I know, no one who is technically literate is an enthusiastic supporter of fuel-cell-powered vehicles," said Donald R. Sadoway, professor of materials engineering and faculty fellow at the Massachusetts Institute of Technology, and a nationally recognized battery expert.

> *'No one who is technically literate is an enthusiastic supporter of fuel-cell-powered vehicles.'*

The assessment of unbiased researchers like Sadoway contrasts sharply with that of automakers, which have cranked up programs to develop the technology, and fuel cell suppliers, which are betting heavily on it. President Bush, too, is a profound believer, which is why he has called for funding of more than $1 billion to support fuel cell research over the next decade. To be sure, many scientists and engineers support the idea of further fuel cell research. But they warn against raising hopes too high, simply because they see no solutions on the horizon for the cost problems, particularly on the automotive side.

"At technical and scientific meetings, we're hearing nothing to lead us to the conclusion that there's been a big scientific breakthrough in fuel cells," said Elton Cairns, professor of chemical engineering at the University of California, Berkeley, and a developer of electric-vehicle batteries for General Motors during the 1970s.

"We need breakthroughs—in electrocatalysis and in polymer exchange membranes—if this vision is going to be realized," Cairns said.

Limited knowledge, high costs

Many researchers fear that hydrogen-powered vehicles will travel the path of the battery-powered vehicle a decade ago. Back then, some suppliers raised great expectations by claiming they could make batteries that would propel cars 400 miles between charges and which would recharge in just 15 minutes. When their research money was used up, however, the claims hadn't materialized. Automakers, meanwhile, have since all but abandoned battery-powered cars.

To avoid repeating that scenario, scientists say the public needs to be made aware that the shortcomings of fuel cells are knowledge-limited, rather than resource-limited. Merely throwing money and other resources at the problem won't guarantee a breakthrough, MIT's Sadoway said.

For example, Sadoway compared the automotive fuel cell dilemma to the search for a cancer cure, another knowledge-limited problem that has remained largely intractable. In contrast, other major scientific efforts, such as the U.S. program to put humans on the moon, were resource-limited. Once the money was there, the job could be done.

"When you have a knowledge-limited problem, there's no assurance you're going to solve it, no matter how hard you try," Sadoway said.

> *Merely throwing money and other resources at the problem won't guarantee a breakthrough.*

The key scientific obstacle facing fuel cell researchers is the fact that automotive fuel cells require large amounts of precious metals—usually platinum—and these are costly. "For fuel cells in automobiles, the issue is cost, cost, cost," said Cairns of UC Berkeley. "And maybe 'cost' a couple of more times."

Indeed, researchers say that fuel cell costs are currently hovering between $1,000 and $3,000 per kilowatt. To compete with vehicles equipped with internal combustion engines, those figures need to plummet to about $30/kW. "As long as you've got to buy your electrodes at the jewelry store, you can bet you're not going to put fuel cells on the road that are competitive with internal combustion engines," Sadoway said.

In products such as laptop computers and cell phones, such costs might be tolerated, scientists say. Laptop owners, for example, have traditionally been willing to pay $5,000/kW to $10,000/kW for batteries. Cell phones often employ batteries that run about $1,000/kW.

That's not to say fuel cells for laptops and cell phones are a slam-dunk, however. Challenges still loom regarding the use of hydrogen and methanol fuels in airports and on aircraft before such technologies can take off in the consumer electronics sector, experts said.

Unquestionably, the problems are much greater in the automotive arena, however. In addition to lower-cost electrode materials, scientists say that advances are needed for the creation of less costly polymer electrolyte membranes (which now run about \$100/square foot) and bipolar plates, which are used between the cells in the fuel stack.

"It would take many grams of precious metals to make a fuel cell stack powerful enough to propel a car," Cairns said. "We're talking about a material demand that would put a strain on the precious-metals market."

Automakers, however, appear undaunted by the obstacles in their path. General Motors Corp. is making a major commitment to the technology, saying that fuel cells provide a design platform unlike anything yet seen in the 100-plus-year history of the automobile. GM is saying little about its technical plans, however. Asked about the need for technology breakthroughs, a spokeswoman said GM engineers know the technical path but declined to elaborate. "We don't want our competitors to know whether we've already achieved it or whether we are still working toward it," she said.

The giant automaker plans to integrate fuel cells with by-wire technologies in a platform that would eliminate transmissions, steering columns, brake linkages, master cylinders and hydraulic fluid lines, enabling engineers to pack all functional components in an 11-inch-deep "skateboard" that would serve as a platform for a variety of vehicle bodies.

Similarly, Ford, DaimlerChrysler, Honda and Nissan are working with Ballard Power Systems Inc. (Vancouver, British Columbia) on fuel cell programs. Virtually all of the automakers are keeping mum on the precise details of their R&D [research and development].

Baffled by carmakers' confidence

General Motors [GM] reinforced its commitment to fuel cells in May 2003, when it announced that Dow Chemical Co. will use 500 of GM's fuel cells to produce up to 35 megawatts of stationary power at a Dow facility in Freeport, Texas. GM executives proclaimed that the goal of the deal was to "reduce the cost of fuel cells and improve their durability so that we may put them in cars by the end of the decade."

Some scientists see value in such efforts. "Car companies are realizing that fuel cells will benefit from the development of stationary applications," said Romesh Kumar, head of fuel cell development in the chemical-engineering department at Argonne National Laboratory. The lab's Transportation Technology Research and Development Center (Argonne, Ill.) is working on various aspects of fuel cell technology, including an on-board vehicle reformer that could convert gasoline to hydrogen for use in an automotive fuel cell power system. "Maybe fuel-cell-powered cars will come at the tail end of those stationary developments," he said.

Still, researchers and battery experts outside the automotive industry say they are baffled by carmakers' confidence in fuel cell technology, especially in light of the industry's battles over battery-powered cars. As recently as two years ago, automakers sued the state of California over a mandate forcing them to sell a designated percentage of "zero-emission" vehicles or face hefty fines. At the time, automotive engineers complained that some battery makers contributed to the problem by overstating the capabilities of their products.

Now, however, it's not the suppliers that are the source of all the optimism, they say. "The auto industry is not getting suckered," Cairns of UC Berkeley said. "They're suckering themselves."

Even Argonne's Kumar, who is a strong proponent of fuel cell research, warned against overoptimism. "People may be willing to pay for the perceived value of a fuel cell in their laptops," he said. "But in an automobile, they may not be."

Many researchers are concerned that the public will misunderstand fuel cell technology and underestimate the difficulties that lie ahead.

"People tend to look down their noses at this kind of technology," MIT's Sadoway said. "But it is every bit as elaborate as anything you'd find in high-energy physics or molecular biology. It's deceptively complex."

"The auto industry is building up public expectations for something that may not materialize," Cairns added. "This simply isn't going to happen unless we see some major breakthroughs."

14

Biodiesel Can Be Used to Heat Homes

Greg Pahl

Greg Pahl has written hundreds of articles and commentaries on the arts, business, finance, farming, wind power, solar energy, electric cars, "green" appliances, home building materials, and sustainable forestry management.

Biodiesel fuel can be made from the oil of soybeans, corn, hemp seeds, or French fry grease taken from fast food restaurants. While it is often used to run cars and trucks, when mixed with fuel oil or used by itself, this clean-burning oil can efficiently heat homes, offices, and other buildings. Biodiesel is easily produced with few environmental consequences and may even be readily available for free at restaurants that have to pay others to dispose of fryer grease. For those who traditionally burn polluting fuel oil for heat, the benefits of biodiesel are many. Biodiesel reduces hydrocarbon and particulate pollution and even reduces the necessity to clean dirty furnaces. By using biodiesel in furnaces, the United States can reduce its dependence on foreign oil and support American farmers who grow the crops used to make the fuel.

Although it has been promoted mostly as a fuel for diesel powered vehicles, biodiesel is perfectly suited as an additive or replacement fuel in a standard oil-fired furnace or boiler.

When used as a heating fuel, biodiesel is sometimes referred to as "biofuel" or "bioheat." Made from new and used vegetable oils or animal fats, this fuel also has the advantage of being biodegradable, nontoxic and renewable: While fossil fuels took millions of years to produce, fuel stocks for biodiesel can be created in just a few months, and the plants grown to make biodiesel naturally balance the carbon dioxide emissions created when the fuel is combusted. What's more, the resulting fuel is far less polluting than its petroleum-based alternative.

The idea of using vegetable oil as a fuel source isn't a new one: In 1900, Rudolph Diesel, a German engineer for whom the diesel engine is named, used peanut oil to power one of his engines at the World Exposition [in Paris].

Today, Rudolph Diesel's original idea of using vegetable oils as a fuel source has been revived with the development of biodiesel.

Technically a fatty acid, methyl ester, biodiesel is made by reacting a wood or grain alcohol such as methanol or ethanol, with vegetable oil or animal fats, with the help of a sodium hydroxide (lye) catalyst, the reaction produces two products: biodiesel and glycerine. The process is relatively simple, although the chemicals required are caustic and need to be handled carefully.

> *I used to refer to biodiesel as an alternative fuel, but now I call it an 'American fuel, made by American farmers.'*

After I first heard about this idea at a renewable energy fair in 2001, I decided to try biodiesel in my old oil furnace. That November, shortly after our fuel tank in the basement had been filled with No. 2 fuel oil, I carefully added about 5 gallons of biodiesel to the tank, which resulted in a B2 blend (2 percent biodiesel; 98 percent fuel-oil).

I started the experiment with such a modest amount because, among its many properties, biodiesel also is a solvent. This potent property tends to dissolve the sludge that often coats the insides of old fuel tanks and fuel lines, which can cause a clogged fuel filter or burner head. As the 2001–2002 heating season progressed, I gradually increased the percentage of biodiesel until the furnace was burning a B10 blend.

Despite my initial concerns, the old oil-fired boiler in the basement continued to operate without any problems. Last year, I increased to a B20 blend, which burned with similar results. . . .

Terry Mason of North Wolcott, Vermont, however . . . started heating his home with B100 about three years ago, making the biodiesel in his basement from recycled cooking oil collected from local restaurants. "I wanted to be self-sufficient in my home heating," he says. "I really didn't have any problems except for a little gunk in the fuel filter the first time I started using the biodiesel."

Great potential

The potential for reducing our reliance on imported crude oil with the increased use of biodiesel as a heating fuel additive is substantial. Officials at the USDA Agricultural Research Center in Beltsville, Maryland, estimate that if everyone in the Northeast used a B5 blend in their oil furnaces, 50 million gallons of regular heating oil a year could be saved.

The Center has been heating its many buildings successfully with a biodiesel blend since 2000. They started by burning a B5 blend, but in 2001, encouraged by the test results, they switched to B20 without experiencing any problems.

"Using biodiesel offers an opportunity to reduce emissions, especially particulate matter and hydrocarbons, and that's a great advantage," says John Van de Vaarst, Agricultural Research Center deputy area director, who is responsible for facilities management and operations. "I used to refer to biodiesel as an alternative fuel, but now I call it an 'American fuel, made by American farmers.' I think it's an obvious strategy to help clean up the environment and reduce our dependency on foreign oil."

Sponsored by the National Renewable Energy Laboratory and the U.S. Department of Energy, Brookhaven National Laboratory on Long Island conducted its own series of tests on the use of biodiesel for space heating.

That facility's 2001 test report found that biodiesel blends at or below B30 can replace fuel oil with no noticeable changes in performance. Burning of the blends also reduced carbon monoxide and nitrogen oxide emissions.

Supply and demand

U.S. biodiesel production was 15 million gallons in 2002 and should reach about 20 million gallons by the end of 2003, according to Jenna Higgins, director of communications for the National Biodiesel Board, headquartered in Jefferson City, Missouri.

"There has been a lot of interest, particularly in the Northeast, in using biodiesel as a home heating oil," she says. "I think it's definitely a very strong potential market in the future."

Roughly three out of four U.S. homes using heating oil are in the Northeast, so the potential for expanding the use of biodiesel in that region is substantial.

> // 'It's a very easy match for home heating, particularly if you have an indoor storage tank. . . . Other than that, there really isn't anything that has to be done in order to use it.' //

Residential consumption of No. 2 heating oil in 2001 was 6.6 billion gallons nationwide, according to the Energy Information Agency.

If every homeowner in this country, currently heating with oil switched to B20, 1.3 billion gallons of biodiesel would be needed. According to the U.S. Department of Energy (DOE), enough feedstocks exist today to produce 1.9 billion gallons of biodiesel. Another 5 to 10 billion gallons could be made from mustard seed, and billions more could potentially be made from algae. U.S. production of biodiesel could climb to 2.5 billion gallons per year by 2020, according to DOE projections.

Real-world tests

Since 2001, the Warwick, Rhode Island, school district also has been conducting biodiesel fuel tests. During the first heating season, the district burned three different percentages of biodiesel (B10, B15 and B20) as well as a No. 2 fuel-oil control in a fourth school.

"It worked very, very well for us," says Bob Cerio, energy manager for the district. "We had three different types of burners, three different types of boilers, and three different sizes; so we had an opportunity to test a wide spectrum. With the smaller boilers, we were able to get similar test data to what people would be experiencing in their home."

After a successful first season, Cerio switched to a B20 blend for the 2002–2003 heating season without any problems. The district continues to use B20 and is no longer experimenting with any lower-percentage blends.

Cerio also tested boiler efficiency and measured emissions. He says although there has been no change in efficiency, emission reductions have been measured in sulfur dioxide, nitrous oxides, carbon monoxide and carbon dioxide. "We've also discovered that our boilers are running much cleaner, so that saves us quite a lot of work cleaning them."

He is enthusiastic about the use of biodiesel as a home heating fuel. "It's a very easy match for home heating, particularly if you have an indoor storage tank," he says. "Other than that, there really isn't anything that has to be done in order to use it.". . .

Catching on

In 2002, Frontier Energy, Inc., of South China, Maine, began to offer biodiesel to homeowners in its regular delivery area, between Augusta, Maine, and Waterville, Maine. The company offers and actively promotes a B5 "Basic Bioheat" blend as well as a B20 "Premium Bioheat" blend. And for those who want it, B100 also is available, although the company doesn't recommend using it as a heating fuel at that concentration.

> *The comment I usually get is, 'I can't tell the difference,' which is exactly what you want to hear.*

"It's going very well so far," says Joel Glatz, vice president for Frontier Energy. "We're probably selling about the same amount for vehicular use as we are for heating use at this point, but I think the heating application is what is really going to catch on in this state. We use about 400 million gallons in Maine for heating oil and about 150 million gallons for transportation annually, so, obviously, there is a much larger market for heating in this state."

Homeowner response has been extremely positive, Glatz says. "Those who have used it, love it. The comment I usually get is, 'I can't tell the difference,' which is exactly what you want to hear."

Another new company to join the biodiesel market is Vermont's Alternative Energy Corporation (VAEC) of Williston, Vermont. Launched in early 2003, the company offers biodiesel heating-oil blends to residential customers through partnerships with large oil companies with terminals in the state.

"We are looking at dealers across the state that we've targeted for distribution," says VAEC president Greg Liebert. In addition to home heating fuel, VAEC also offers biodiesel for vehicles, and biodiesel processing equipment, education and training. In the near future, the company hopes to establish commercial-scale biodiesel processing facilities.

In some parts of the country, homeowners who have been frustrated by the lack of local distributors have formed energy co-ops, through which they order biodiesel in large quantities and at lower prices. Co-opPlus, a member owned energy cooperative in western Massachusetts, is involved in a variety of renewable energy programs, including a biodiesel initiative that now is associated with Alliance Energy Services in Holyoke, Massachusetts. Alliance currently offers a B20 blend as well as B100.

"Biofuel is readily available, and it makes sense for a lot of people," says Stephan Chase, the company's president. Alliance, which has been actively promoting its biofuel, has about 100 biodiesel customers and a growing demand for biofuel. "It will be interesting to see what happens," Chase says. "The biofuel is a good product, and the Pioneer Valley [in western Massachusetts] has a lot of residents who are concerned about the environment, so it's a good combination; we should do very well with it here."

Future prospects

Biodiesel is a simple, proven fuel that, along with other renewable fuels and conservation strategies, could help end U.S. dependence on foreign crude oil and dramatically improve air quality nationwide.

"It has the capability of giving our farmers a good, steady cash crop, helping our economy, reducing our dependency on the foreign oil market, and it's the right thing to do for the environment," says the New Jersey school district spokesman Cerio, "and it's far beyond the experimentation phase at this point."

If you already heat with oil, can find a local supplier and are willing to pay a little more, using biodiesel will let you stay warm this winter in a much greener way.

Organizations to Contact

The editors have compiled the following list of organizations concerned with issues debated in this book. The descriptions are derived from materials provided by the organizations. All have publications or information available for interested readers. The list was compiled on the date of publication of the present volume; the information provided here may change. Be aware that many organizations take several weeks or longer to respond to inquiries, so allow as much time as possible.

American Wind Energy Association (AWEA)
122 C St. NW, Suite 380, Washington, DC 20001
(202) 383-2500 • fax: (202) 383-2505
e-mail: windmail@awea.org • Web site: www.awea.org

The American Wind Energy Association is a national trade association that represents wind power plant developers, wind turbine manufacturers, utilities, consultants, insurers, financiers, researchers, and others involved in the wind industry. The AWEA promotes wind energy as a clean source of electricity for consumers around the world. The association provides statistics and information on development in the domestic and international markets to industry interests, the general public, and the news media. Publications include industry documents, the latest economic studies, technology information, and books.

Committee for a Constructive Tomorrow (CFACT)
PO Box 65722, Washington, DC 20035
(202) 429-2737
e-mail: info@cfact.org • Web site: www.cfact.org

The Committee for a Constructive Tomorrow supports continued development of technologies such as agricultural chemicals, atomic power, and biotechnology. CFACT works to promote free market and technological solutions to such growing concerns as energy production, waste management, food production and processing, air and water quality, and wildlife protection. The committee produces a national radio commentary called "Just the Facts" that is heard daily on about three hundred stations across the country.

Competitive Enterprise Institute (CEI)
1001 Connecticut Ave. NW, Suite 1250, Washington, DC 20036
(202) 331-1010 • fax: (202) 331-0640
e-mail: info@cei.org • Web site: www.cei.org

CEI encourages the use of the free market and private property rights to protect the environment. It advocates removing governmental regulatory barriers and establishing a system in which the private sector would be responsible for the environment. CEI's publications include the monthly newsletter *CEI Update* and editorials in its On Point series, such as "Property Owners Deserve Equal Access to Justice."

Ecological Life Systems Institute (ELSI)
Ecological Life Systems Institute, Inc.
PO Box 7453, San Diego, CA 92167
(619) 758-9020 • fax: (619) 758-9029
e-mail: info@elsi.org • Web site: www.elsi.org

The goals of the Ecological Life Systems Institute include learning how people can earn a living in ways that are economically sustainable. To do so, the ELSI identifies and promotes policies and practices that advance the health of the environment and teaches that knowledge to others in classrooms, industry, and government. The institute's Web site has dozens of articles concerning renewable energy, global warming, and related topics.

EnviroHealthAction (EHA)
1875 Connecticut Ave. NW, Suite 1012, Washington, DC 20009
(202) 667-4260
e-mail: info@envirohealthaction.org
Web site: www.envirohealthaction.org

EnviroHealthAction is an education and advocacy center sponsored by Physicians for Social Responsibility. EHA provides information about the negative effects of pollution on human beings to health care professionals and others interested in environmental health. It is designed to provide professionals with the opportunity to access important resources and deliver well-researched input to policy makers. EHA publishes research papers on toxics and health, children's environmental health, air pollution and health, chronic disease and the environment, safe drinking water, and other issues.

Heartland Institute
19 S. LaSalle St., Suite 903, Chicago, IL 60603
(312) 377-4000
e-mail: think@heartland.org • Web site: www.heartland.org/Index.cfm

The Heartland Institute is a conservative source of research and commentary. Heartland's mission is to help build support for ideas that include parental choice in education, choice and personal responsibility in health care, market-based approaches to environmental protection, privatization of public services, and deregulation. The institute publishes a monthly online newsletter on environmental and other issues.

The Heritage Foundation
214 Massachusetts Ave. NE, Washington, DC 20002
(800) 544-4843 • fax: (202) 544-2260
e-mail: pubs@heritage.org • Web site: www.heritage.org

The Heritage Foundation is a conservative think tank that supports the principles of free enterprise and limited government in environmental matters. Its many publications include the following position papers: "Can No One Stop the EPA?," "How to Talk About Property Rights: Why Protecting Property Rights Benefits All Americans," and "How to Help the Environment Without Destroying Jobs."

National Biodiesel Board (NBB)
3337A Emerald Ln., PO Box 104898, Jefferson City, MO 65110
(573) 635-3893 • fax: (573) 635-7913
e-mail: info@biodiesel.org • Web site: www.biodiesel.org

The National Biodiesel Board is a national trade association representing the biodiesel industry and acts as the coordinating body for biodiesel research and development in the United States. It was founded in 1992 by state soybean commodity groups, who were funding biodiesel research and development programs. Since that time, NBB's membership has been comprised of state, national, and international feedstock and feedstock processor organizations, biodiesel suppliers, fuel marketers and distributors, and technology providers. In addition to distributing books and brochures concerning biodiesel the NBB publishes the *Biodiesel Bulletin*, a monthly newsletter.

The National Renewable Energy Laboratory (NREL)
1617 Cole Blvd., Golden, CO 80401-3393
(303) 275-3000
Web site: www.nrel.gov

The National Renewable Energy Laboratory is the U.S. Department of Energy's laboratory for renewable energy research, development, and deployment, and a leading laboratory for energy efficiency. The laboratory's mission is to develop renewable energy and energy efficiency technologies and practices, advance related science and engineering, and transfer knowledge and innovations to address the nation's energy and environmental goals. Some of the areas of scientific investigation at NREL include wind energy, biomass-derived fuels, advanced vehicles, solar manufacturing, hydrogen fuel cells, and waste-to-energy technologies. The organization publishes dozens of comprehensive research papers concerning these technologies, many of them available for free online.

Natural Resources Defense Council (NRDC)
40 W. Twentieth St., New York, NY 10011
(212) 727-2700 • fax: (212) 727-1773
e-mail: nrdcinfo@nrdc.org • Web site: www.nrdc.org/default.asp

Natural Resources Defense Council is one of the nation's largest environmental action organizations. It uses law, science, and the support of more than a million members and online activists to protect the planet's wildlife and wild places. The NRDC works to restore air, land, and water and to defend endangered natural places.

Nuclear Energy Institute (NEI)
1776 I St. NW, Suite 400, Washington, DC 20006-3708
(202) 739-8000 • fax: (202) 785-4019
e-mail: webmasterp@nei.org • Web site: www.nei.org

The Nuclear Energy Institute is the policy organization of the nuclear energy industry whose objective is to promote policies that benefit the nuclear energy business. NEI develops policy on key legislative and regulatory issues affecting the nuclear industry. The organization has over 260 corporate members in fifteen countries, including companies that operate nuclear power plants, design and engineering firms, fuel suppliers and service companies, and companies involved in nuclear medicine and nuclear industrial applications. NEI publishes numerous books and brochures that promote nuclear energy and safety.

Renewable Energy Policy Project (REPP)
1612 K St. NW, Suite 202, Washington, DC 20006
(202) 293-2898 • fax: (202) 298-5857
e-mail: info2@repp.org • Web site: www.repp.org

Renewable Energy Policy Project provides information about solar, hydrogen, biomass, wind, hydrogen, and other forms of "green" energy. The goal of the group is to accelerate the use of renewable energy by providing credible facts, policy analysis, and innovative strategies concerning renewables. REPP seeks to define growth strategies for renewables that respond to competitive energy markets and environmental needs. The project has a comprehensive online library of publications dedicated to these issues.

Sierra Club
85 Second St., Second Fl., San Francisco, CA 94105-3441
(415) 977-5500 • fax: (415) 977-5799
e-mail: information@sierraclub.org • Web site: www.sierraclub.org

The Sierra Club is a nonprofit public interest organization that promotes conservation of the natural environment by influencing public policy decisions—legislative, administrative, legal, and electoral. It publishes *Sierra* magazine as well as books on the environment.

The Union of Concerned Scientists (UCS)
National Headquarters, 2 Brattle Square, Cambridge, MA 02238-9105
(617) 547-5552 • fax: (617) 864-9405
Web site: www.ucsusa.org

UCS is an independent nonprofit alliance of more than one hundred thousand citizens and scientists who link scientific analysis with citizen advocacy to build a cleaner environment and a safer world. UCS was founded in 1969 by faculty members and students at the Massachusetts Institute of Technology who were concerned about the negative effects of science and technology in society. UCS publishes numerous books and articles concerning food safety, clean vehicles, global environment, clean energy, and global security along with several periodicals including *Catalyst, Earthwise, Gene Exchange*, and *Nucleus*.

Worldwatch Institute
1776 Massachusetts Ave. NW, Washington, DC 20036-1904
(202) 452-1999 • fax: (202) 296-7365
e-mail: worldwatch@worldwatch.org • Web site: www.worldwatch.org

The Worldwatch Institute is a research organization that focuses on key environmental, social, and economic trends. Its goal is to promote a transition to an environmentally sustainable and fairly run society. The institute is responsible for publishing many books and publications, including the annuals *State of the World* and *Vital Signs* and the bimonthly *World Watch* magazine.

Bibliography

Books

Rick Abraham — *The Dirty Truth: George W. Bush's Oil and Chemical Dependency: How He Sold Out Texans and the Environment to Big Business Polluters.* Houston: Mainstream, 2000.

Paula Berinstein — *Alternative Energy: Facts, Statistics, and Issues.* Westport, CT: Oryx, 2001.

Anne-Marie Borbely and Jan F. Kreider, eds. — *Distributed Generation: The Power Paradigm for the New Millennium.* Boca Raton, FL: CRC, 2001.

Edward S. Cassedy — *Prospects for Sustainable Energy: A Critical Assessment.* New York: Cambridge University Press, 2000.

Seth Dunn — *Hydrogen Futures: Toward a Sustainable Energy System.* Washington, DC: Worldwatch Institute, 2001.

Ursula Eicker — *Solar Technologies for Buildings.* Hoboken, NJ: Wiley, 2003.

Rex A. Ewing — *Power with Nature: Solar and Wind Energy Demystified.* Masonville, CO: PixyJack, 2003.

Howard S. Geller — *Energy Revolution: Policies for a Sustainable Future.* Washington, DC: Island, 2003.

Richard Heinberg — *The Party's Over: Oil, War, and the Fate of Industrial Societies.* Gabriola, BC: New Society, 2003.

Geoffrey Frederick Hewitt — *Introduction to Nuclear Power.* New York: Taylor & Francis, 2000.

Peter Hoffmann — *Tomorrow's Energy: Hydrogen, Fuel Cells, and the Prospects for a Cleaner Planet.* Cambridge, MA: MIT Press, 2001.

Michael Potts — *The New Independent Home: People and Houses That Harvest the Sun.* White River Junction, VT: Chelsea Green, 1999.

Periodicals

Alternative Fuels Data Center — "Alternative Fuels—Information and Comparisons," www.afdc.nrel.gov/altfuels.html.

Lewis Braham — "Alternative Energy: Want the Last Laugh? While Wall Street Sneers, Savvy Investors May Find Opportunities," *BusinessWeek*, June 2, 2003.

Ty Cashman — "The Hydrogen Economy," *Earth Island Journal*, Summer 2001.

Ann Chambers — "Wind Project Siting Faces Unique Hurdles," *Power Engineering*, May 2000.

Mona Chiang — "Recharge! In a Polluted World, Can Nature Provide Cleaner Energy?" *Science World*, April 18, 2003.

Lyn Collins — "Renewable Energy," *Geography Review*, September 2001.

James Detar — "Fuel Cells May Power Up Computer Sales; Running Portable Gear Longer; Hydrogen Fuel Cells Act Like Batteries, but Don't Run Down or Require Recharging," *Investor's Business Daily*, September 9, 2003.

Manimoli Dinesh — "Hydrogen Fuel No Near-Term Panacea for Oil Dependence," *Oil Daily*, February 9, 2004.

Dave Duffy — "How Environmental Ideology Hurts the Solar Energy Industry," *Backwoods Home Magazine*, May 2001.

Energy Intelligence Group — "Big Oil Role in Fuel Cells Looks Secure," *Petroleum Intelligence Weekly*, March 3, 2003.

Glenn Hamer — "Tipping Point: View on Renewables," *Power Engineering*, May 2003.

Hempcar Transamerica — "Pollution: Petrol vs. Hemp," November 2000. www.hempcar.org/petvshemp.shtml.

Junona Jonas — "Primer on Geothermal Energy," *Electric Light & Power*, November 2003.

Lynn Osborn — "Energy Farming in America," Hempcar Transamerica, 2000. www.hempcar.org/efia.shtml.

Alan Reder — "Here Comes the Sun: Solar Power, Once the Energy Source of Choice for Hippies Living in Backwoods Shacks, Has Finally Made It to the Suburbs. Hey, Man, Now You Can Have It All," *OnEarth*, Winter 2004.

Paul Roberts — "Running Out of Oil—and Time," *Los Angeles Times*, March 7, 2004.

Janet L. Sawin — "Run with the Wind," *New Internationalist*, June 2003.

George Sterzinger — "OPEC and U.S. Energy," *Harvard International Review*, Winter 2004.

Joe Truini — "A Mighty Wind: Renewable Energy Is Becoming More Price-Competitive," *Waste News*, June 23, 2003.

Paul Weinberger — "Balancing Potential with Limitations," *Power Engineering*, March 2003.

Index